# A Beginner's View of Our Electric Universe

By Tom Findlay

*"We should not surrender our judgement to others; we must reclaim our ability to doubt and think for ourselves."*

Tom Findlay

This book is published by
Grosvenor House Publishing Ltd
28-30 High Street, Guildford, Surrey, GU1 3EL.
www.grosvenorhousepublishing.co.uk



A CIP record for this book
is available from the British Library

ISBN 978-1-78148-141-7

# Contents

## Reviewer comments

“Tom Findlay's book is a remarkable contribution from a newcomer to the Electric Universe. He shows the impact that this new and simpler way of seeing the universe can have on a practical man with a keen interest in astronomy. "A Beginner's View" is easy to read and copiously illustrated. Tom makes a heartfelt plea for individuals to participate in science once more; to use their intuition and common-sense to question the science fiction headlines and gross expenditure on massive projects. After all, history shows most great breakthroughs are made by individuals, most of them outsiders.”

**Wallace Thornhill,** lead author and researcher of **“The Electric Universe”**

“Newcomers to the idea of electricity in space are apt to find even the possibility of electrical explanations for cosmic phenomena to be unfamiliar and therefore startling. This book provides a comfortable transition from familiar gravity-based explanations to a basic understanding of plasma behavior and its manifestations in stars and galaxies. The book will ease the surprise and enable the reader to better understand the more technical publications in this new way of thinking about the universe.”

**Meldon Acheson**

“One of the biggest challenges facing modern science is ensuring that the members of the general public fully understand what has been achieved in all areas of science and what is proposed for future research. This is vital because it is that public which eventually pays the bill for the 'games scientists play'. The advocates of certain theories have been eminently successful in this enterprise but the end result has been that the public has been left woefully ignorant of many of the very real controversies existing. This is especially true in the general area of astronomy/astrophysics. For a great many years now, that general area has been dominated by theories which elevate the force of gravity to a dominant position, while ignoring the possible effects of the much much stronger electromagnetic force. Workers in the field covered by the names plasma cosmology and electric universe have attempted to rectify this position. However, most of the existing material in this area is hardly in a form for consumption by the general public. This present book by Tom Findlay, “A Beginner’s View of Our Electric Universe”, is an excellent attempt to rectify this situation. In it, readers will find explanations for a great number of astrophysical phenomena in terms of electric universe ideas; some even clear up situations not understood on the basis of currently accepted conventional wisdom. Although in a sense a 'popular' science book, this book will require the reader to concentrate if a real understanding is the requirement but that concentration will be very well rewarded at the end and the reader will realise just how much more there is to learn about our solar system, our galaxy and even our universe than is portrayed in present 'pop-science' books in the area. I urge all to read this book, digest its contents and finish with a fuller, more complete understanding of our wonderful universe that will open the door for delving deeper into the greater technical detail available in books and on websites published and maintained by the researchers at the core of this topic."

**Dr Jeremy Dunning-Davies,** Hull University Physics Department (retired) and author of **"Exploding a Myth"**

"Reading this book left me with the impression that it could be considered encyclopedic in its breadth of description of the elements of science and observation that are basic to the Electric Universe paradigm. By addressing the most interesting and provocative detail of the Electric Universe model, this book covers the bases that lead to the more technical "Essential Guide to the Electric Universe". If this straightforward read doesn't get lay people as well as astronomers and astro-scientists to thinking about what today's observations might really mean, and how to effectively incorporate better explanations into the fabric of the science being taught to our students all over the world, then it's hard to imagine what will. The illustrations spread throughout are a valuable element, for we all see and understand much better when details can be explained graphically. The NASA images, without much further discussion, or, too often, ill-conceived conjecture based on old assumptions, say much more than words alone. Tom's own illustrations are clear, artistically done, and convey the appropriate fundamentals of electricity and electrodynamics in a way almost anyone can pick up. The book's organization is clear and concise, and follows a logical path from beginning to end. The nod that is given to the scientists, whom history is apparently trying to relegate to the dust bin, is also appreciated. It is true that unfortunate things have happened as good science has been ignored in the past, but change is now afoot, and this book is an example of the rash of material now appearing, which by presenting relevant elements of the history of scientific revolutions and the discoveries that have arisen from them, opens the door to the inescapable realization that the old order in astro-science will be replaced in a generation or two. This will be the first introduction to the Electric Universe for a lot of people. I think that for many it will be the start of a lifelong change of perspective and knowledge about what kind of Universe we are an intrinsic part of."

**Jim Johnson -** Electric Universe researcher and editor of **"The Essential Guide to the Electric Universe"**

"In general I feel strongly that this book will serve nicely as another brick in the wall of evidence that will be convincing to the average reader. Anyone who has interest enough to read it cannot go away without having the feeling that: the EU has a great deal to offer modern astro-science; the power structure of modern astro-science is reacting irrationally to our suggestions; an honest, intelligent man who does not pretend to have expertise in astro-science, but who does have a deep interest in it wrote this work. The author also clearly has the ability to incisively ferret out subtle errors as well as obvious nonsense arguments and baseless assumptions. The style is very appealing."

**Dr Donald E Scott** - Professor of Electrical & Computer Engineering at the University of Massachusetts/ Amherst (retired) and author of "**The Electric Sky**".

# Dedication

***For my wife Nora, my sweetheart and best friend and for our two great sons, Ben and Jon.***

## Acknowledgements

I claim no originality for any aspect of science theory in this book. All I have done is attempt to produce my own interpretation of the work of experts in these subjects. The people from whose work I have drawn and those who have helped me significantly along the way include:

***David Talbott and Wallace Thornhill*** – Dave and Wal, my admiration for your work and dedication to this subject is total. What I have learned from your individual and combined works has provided a new understanding that has allowed me to see further and clearer than my previous horizon. Dave, you saw my eagerness and perhaps some potential so you supported me and opened doors that I would never otherwise have known existed. Wal, your clear descriptions of your own theories and those of others have been my most significant resource as I built my big picture of the science that underpins the Electric Universe model. The book that you and Dave produced together, '***The Electric Universe***', is always close to hand.

I owe you another debt, Wal. Although your time is precious and demanded by so many people and projects, you gave it willingly to review and provide comprehensive feedback on all I had written. This was a significant effort and contribution that I appreciate deeply. What can I say here other than thank you, most sincerely.

***Donald E Scott*** – Don, we never met, I truly wish we had. You appear the type of person that likes to get straight to the point and I appreciated that as I read and learned so much from your book '***The Electric Sky***'. It is also a reference source that is never far away. Thank you also for reviewing and suggesting improvements to my book; I took much needed energy from that and will remember fondly your words of encouragement.

***Mel Acheson*** – Again, Mel, we never met but I really wish we had. Twice I asked for your help in reviewing my work; once for a conference paper and once again for my book. You willingly gave your time and expertise to a detailed review of both and I thank you most sincerely for your time and expertise in doing that. You too had encouraging words for me that helped provide necessary momentum at the right time. I will not forget.

***Jim Johnson, Jeremy Dunning-Davies, Gerard Bik*** and ***A.P. David*** ... What a great bunch of guys. You all unconditionally gave me your time by reviewing my work as it was slowly being produced then supplied from your own expertise the feedback that enabled me to pat each piece into an initially acceptable shape. I truly have no clear idea where things would have gone without the involvement of all of you good people.

***Jim***, you, just like our good friend ***Michael Gmirkin***, are no less that an amazing walking encyclopaedia. Thank you so much for giving me the benefit of your impressive knowledge and infectious, inquisitive mind!

***Jeremy***, I knew, after I got to know you a little and after reading your own book '***Exploding a Myth***' that I was going to be dealing with a wise and experienced head that would provide advice I could bank on. Thank you so much for your expertise, sensitive guidance and timely encouragements.

***Gerard***, your eye for detail has been a joy to learn from and your courteous, prompt responses have allowed me to clarify and add many polishing touches to my work. I appreciate greatly the patience you have shown and the obvious wealth of knowledge from which you have drawn.

***David***, your eloquent down-to-earth commentary and timely encouraging words have also meant so much to me. The attention you paid to my writing and the advice you gave reinforced my confidence at a particular low point. Thank you; I hope we have the chance to meet again.

***Steve Smith*** – Steve, I have learned so much from your brilliant contributions to the thunderbolts.info website's TPOD (Thunderbolts Picture Of the Day) article collection. I greatly admire your knowledge and that of the other occasional contributors to the TPODs. It was my honour and privilege to meet you and Tiffany in person at the NPA18 conference in 2011.

***Dave Smith***. G'day mate! ... The devotion of your substantial skill set to spreading the word about the Electric Universe is nothing short of totally admirable. Thanks for the long discussions on Skype in the wee small hours that helped give me confidence and for doing the work to make this book available for download on the web. I hope to have the pleasure of buying you that pint some day!

***Marcus Bowman*** – My friend, you read and commented on the whole book when it was in its early stages and provided guidance from your substantial literary and technical expertise on what you found. The book and my approach to its production benefited greatly from that advice and I thank you so much for that.

***Geoff Harland*** – Your contribution through a total review of my work in its latter stages brought many improvements. I was surprised to see you picking up on the simplest of errors that previous eyes had missed or where I had re-introduced them. Your commentary and questions on the technical content and sentence construction made me develop, I hope, a better way of saying things. Thank you, I am most grateful.

***Tony Fyler*** – When the time eventually came, I was super impressed by the professional editing service you supplied. Your company, Jefferson Franklin Editing (UK), lived up to all the promises it made, and more. I thank you sincerely Tony for your honest critique, genuine interest and prompt service.

There are other people who have helped me that I could mention, most of whom are closely associated with '***The Thunderbolts Project***'. They know themselves who they are because they have provided occasional help to me during this project. My sincere thanks to all you good and knowledgeable people.

Most importantly, I must mention those in my daily life who have believed in my ability to pull this off. They took time to understand the job I had set myself and continually provided me with encouragement that helped keep my nose to the grindstone. Some of these dear people are:

My wonderful sister, ***Jean*** – Thanks Jeannie! No other person apart from the two of us will ever know how your patience, practical support and professional guidance has helped turn this book into a reality. You are a sparkling gem of goodness and selfless giving, my dear sister.

My brother, ***Jim*** – Thanks Jimmy for your helpful comments when you saw that I needed a boost. Listening to your thoughts, forged from hard won experience of the world, helped me view anew how I should consider the overall writing process. You've been a real help, buddy!

My '***Toolbox***' – I have never seen it done before, but I just have to say a word or two about my trusty old pal here, my computer. I have continually assured it over the past many months that I appreciate its almost faultless performance and I have nurtured it with all appropriate software checks and updates to ensure that its good behaviour would continue. It has rewarded me a thousand fold, so a big ***thank you*** to my trusty friend, I really do appreciate that you stuck with me!

***Ayr Amateur Radio Group*** – I have been a member of this club for almost 40 years so my friends there just had to be the guinea-pigs first exposed to my thoughts on the Electric Universe model. They sat patiently for many hours listening to me prattling on about it, and it was as a result of the rapid up-take and positive reaction they displayed that the idea to provide an introduction to the Electric Universe for others first came about. So, to this bunch of friends, a couple of whom remain lovable sceptics, I extend my sincere thanks and appreciation.

My deepest appreciation goes to my darling wife Nora for her patience in all matters, loving attention and long periods of help with proof reading. Without her unconditional support this book would never have seen the light of day. Nora, my much better half, you have my undying love, respect and admiration.

# Foreword

I continue to be amazed at how easily we accept whatever the media presents to us. We seem especially not to think too much about what we are told by various high-profile organisations, institutions and people for whom we have some form of respect. In many cases this is fine, but in others, if we had gone to the bother of forming a reasoned opinion of what had been said, we would likely have come up with legitimate questions. The greatest example of this, for me, is the information coming from ***theoretical astro-science*** and the apparently confident air with which it is presented to the public. As I intend to describe, I see this as a particular issue, in that we have traditionally tended to automatically accept as correct what we are told by so-called experts - the folks whom we just assume are on the right track.

I decided to write this book because I wanted to provide an opportunity for people like me to think a bit more about the information presented by astro-science. My approach has been to try to keep things straightforward and not introduce language that might put an interested reader off. I have included a good number of what I think are uncomplicated diagrams together with photographs and other images to support what I describe. From start to finish the content is intended to have a logical flow and to make sense as an introduction to a very large subject area that I do not have the knowledge and skill to address in full detail. I invite you therefore to view this offering as a doorway, one which I hope you go through to find a more detailed, incredibly fascinating and challenging body of information that has relevance for us all.

To those of you who know much more than I do about these subjects, I apologise in advance for the summary nature, absence of supporting scientific descriptions and copious references to peer-reviewed sources – these things are not necessary for the purpose of this introductory book. I feel the most important thing is to engage the interest of members of the public who are motivated to understand more, some of whom may already have suspicions about what astro-science has been telling us.

This is not a complete guide to how our universe works, for that is far too broad and deep a subject for me ever to tackle. Instead, what I offer is a readable summary that restricts itself only to major aspects of theories that support two different points of view. This is necessary for the objectives of the book, two of these being: to turn a spotlight on assumptions that lie behind the story of our universe that we have already been given by astro-science, 'a story now judged by many as incomplete or incorrect,' and then; to describe a much more obvious, sensible and logical set of theories that are making great headway in terms of providing clarity for those who study them. I will expand on these theories not just through my own opinion but from the work of highly qualified scientists, science authors and independent researchers. In doing this I hope to educate, entertain and surprise the reader, and of course, provide food for further thought.

In terms of my personal motivation to write this book, and while again recognising the spark lit for me by my radio club friends as I mentioned in my acknowledgements, I have always been a supporter of what I perceive to be just causes. The strong desire to see good and fair things done and said has been a personal driving force for as many years as I can remember. On occasion, I have felt the need to get personally involved due to some particular manifestation of unfairness or inappropriate behaviour. As you will come to see, this is one such case. However, this motivation does not stand on its own, for I am also intensely interested, as a hobby subject, in how our universe really works and what everything in it is about.

As for my credentials, I have no academic achievements to parade for anyone to inspect but I do, through my career, have knowledge and experience of electrical, electronic and computer engineering, a measure of common sense and an ability to think logically. Ironically, perhaps, the absence of academic moulding may have left me with an ability to fit in place, quite easily, many of the clues we are exposed to today as our awareness of the world and the universe develops. Technical hobbies and group projects of a practical nature have always been more to my taste than involvement with large crowds or a need to maintain a physical competitive edge. One such area of interest gripped me strongly when I was very young; this being the subject of astronomy. I remember listening to the comments my father would make as he gave his opinion to our black and white TV set when Patrick Moore (eventually to become Sir Patrick) of 'The Sky at Night' programme fame, was explaining from his own viewpoint the objects and events we could observe in those days in our heavens. Since then, a lack of funds and cold and cloudy nights here in Scotland have steered my interest in astronomy towards the intellectual goal of understanding what I can about the workings of our universe. Looking back, this could be considered the low-cost option, because learning from the work of others through books does not need to come with the additional expense of buying equipment. This change of emphasis was also the start of my journey into cosmology, physics and some other related areas. As a period of learning, this was ultimately to unfold in ways that I could never have predicted. So began the journey that has now led me to the realisation that there is a very big and very important hidden story to be told, and that if we are to advance our knowledge through better ways of doing science, then there are some fundamental changes that need to be made.

As individuals, I think we feel an inner drive to understand more about things around us but find ourselves sometimes put off by an impression of complexity. I have tried not to see this as a barrier and so have worked to develop a basic understanding of the theories I write about here. I am now motivated to pass on my thoughts about how, through surprisingly common theories, everything in this amazingly dynamic universe can be better explained than it currently is. Additionally, I want to highlight that 'we the public' must pay more attention to science, and where it is correct to do so, we should question the quality and integrity of information provided to us. What I have written is not intended as a condemnation of today's science; rather, it is offered to complement what we already have and to act as a wake-up call to the public to talk more about these things. Please understand as you carry on that I am speaking as an individual through a heart-felt belief in the accuracy of what you will come here to know as the Electric Universe model.

And so why should the theories behind the Electric Universe interest you and why might you want to read this book? To this I feel there are two major responses; one that honours the pursuit of accurate information and another that I consider to be a moral imperative. When we come across important information which through qualified sources and our own best judgement we learn to be incomplete or even faulty, we must highlight this and seek the opinion of others more knowledgable in order to correct the detail. And if, through our studies and research we come across behaviour that is plainly wrong and nothing is apparently being done about it, then we should consider highlighting this and promoting that which in opposition is right and true. So the reason people should read this book is simply for their own education, independent of mainstream thinking, in the pursuit of better scientific information produced through work based on clear moral standards.

---

As you progress you will likely find, especially in the early chapters, the use of some words and descriptions that are new to you. If this is the case, then hanging on till chapter four or searching on the Internet might help. In any case, I hope you stick with it and are not put off. Things will become clearer as you progress. You will also find embedded references to external information formatted as shown here: [x-y] [x = chapter number - y = reference number]. All of these coded links point to text entries at the end of the book in the reference section and in the .pdf version when read on a computer with an online connection, to external web sources.

Please note in particular:

By producing this book, I in no way claim association with the people whose work I refer to and from which I have formed my analysis and opinions. The book is therefore all my own work, so any comments or criticism it provokes should be directed at me and no other person. A feedback form for this purpose has been made available at www.newtoeu.com

Tom Findlay

# Introduction

*"We have to discard 'modern' physics and return to classical physics of a century ago. This, perhaps, is the greatest hurdle – to discard our training and prejudices and to approach the problem with a beginner's mind."*

(Wallace Thornhill (2006) commenting on the direction astro-science has taken.)

---

We exist in a universe that appears to us on a daily basis to be in balance; it is not apparent that anything actually happens out there on the time-scales we have patience for. The forces we refer to that govern the universe are those we call nuclear, electrical and gravitational. These are the forces which somehow are associated with each other in ways that present to us everything our senses tell us is real. Together and individually, these forces are fundamental to our understanding of how our universe works.

Today, the science behind the workings of our universe is described through the mainstream activities of mathematicians and astro-scientists. This work generally comes under the labels of mathematics, astronomy, astro-science, astro-physics and cosmology, or these same labels with the word 'theoretical' stuck in front. It is confusing to be presented with a variety like this, so for the sake of simplicity in this book I will refer to all possible permutations by the single term ***astro-science***.

If we then consider the range of activities that today's astro-science encompasses, is there a chance that a single message, gathered from across all those areas of work, would also be confusing? Yes, a big chance, and this confusion actually exists. Here, it seems that past attempts by mainstream astro-science to present a coherent story of our universe in a sound scientific way have not met with success. There are two main reasons for this; one, the theories relied upon are not proven and are regarded by many as unsound; two, the perceived complexity of the information presented is off-putting for people trying to understand. Considering the latter aspect alone, the interested public's view of astro-science is therefore not one of a user-friendly subject area. It should therefore not be a surprise to anyone that many of us choose not to pay much attention to astro-science, so we just let it get on with its own business. This situation is further compounded by the 'letting out' in public of certain astro-scientists who like to be seen and listened to and who are particularly attracted to the confusing fine detail and mathematics of their subjects. This introduces further unnecessary complication in the minds of those who listen, and the situation it encourages actually works against all forms of effective public communications. In mentioning this, I fully recognise there are many highly qualified people in science and technology who do a first-class job of communicating their excellent ideas. I will say more about this 'perceived complexity' point.

After having spent a good deal of time studying astro-science as a non-professional, I feel that I can say something about the core theories involved. This is that they are typically understandable for the average interested person, so there really is no need for a science or mathematics background in order to achieve a good appreciation of these theories. Some of them do understandably take a little effort to digest, but most of them can be found clearly described and so are quite easy to understand. Admittedly, there are a few that will only ever make sense within the minds and imaginations of the people who thought them up in the first place. The last group is not covered in this book so we will start by first considering the collection of astro-science theories that are said these days to describe how our universe works and go on from there. *It is very important to note that these standard theories are constantly and casually promoted as accepted fact within the public domain because they are never compared against anything else.* They have their foundations in the two major areas of gravity and relativity, which when taken together provide the basis for the Standard Model (SM) [1-1] of the universe.

Even in my own 'outside science' position, questions arose as I looked into these theories. However, not having a substantial academic grounding in astro-science I could only see these questions as points of note at best and serious logical doubts at worst. The culmination of reading many books on cosmology, relativity, time, and quantum theory left me feeling rather uncomfortable, confused, and significantly dissatisfied. I had been substantially put off by the apparent complexity of it all, so I took this on the chin and initially put it down to my own lack of detailed scientific understanding - something I hoped could be worked on at a later date. The main problems had been with finding and understanding the links between individual theories and in coming to terms with the ever-increasing number of mathematical equations that seemed to conveniently appear in support of explanations. Even though the learning exercise itself was invaluable, it was not a very satisfying experience so I started to lose interest and energy for pursuing the clearer picture I once thought I could obtain. I also formed a very clear opinion of why others might be put off.

Time passed and I did not make any headway. Then, quite by chance, I came across a YouTube video [1-2] that outlined an alternative set of theories on how our universe works. These new theories were described, not in terms of the force of gravity as the main player, but with that role being filled by the force known as electromagnetism (EM). I was pleasantly surprised by my rapid uptake of these new ideas and found myself thoroughly immersed in a description that seemed to be making much more sense than the theories of the SM. It so happened that this new model also fitted well with what I already understood about electricity and magnetism through my work as an electrical engineer and my hobbies of amateur radio and electronics. It all came together as a new paradigm that promised to present clear and comprehensive information right from the start, but it was also obvious that the theories involved were poles apart from those of the SM. As a result, and due to the profound implications these radically different theories presented, I found myself on many occasions staring blankly at the wall as the hands on my desktop clock crept round. I was truly lost in contemplation, especially about the serious implications of these new theories being true.

As a memorable process in itself, this took a few days to get through but I eventually found myself coming to terms with my new-found view of things. I had come to understand and accept the fundamental differences between the gravity-based theories of the SM and the electromagnetic theories of this new Electric Universe (EU) model. However, the very much more challenging aspect of this had been to deal with the implications of all this being correct. Thinking about these left a deep impression because they seemed to be very broad and fundamental for everybody.

It struck me in particular that if EU theory was indeed the way things work then it would falsify a great deal of what we have been led to believe and are still being told in regard to how our universe works. It meant that certain theories studied by students in schools, colleges and universities were knowingly wrong at worst and misguidedly convincing at best. It meant too, that much of the scientific work that has gone on and still goes on, together with the funding it has soaked up through public taxation, would have made little real contribution and been of questionable value. I thought also about the careers of those who had dedicated their lives to scientific research based on the gravity and relativity pillars of the SM. I could not help but think they might have wasted their time … This sure seemed crazy stuff!

But hold on a minute, how could I as a layman in these matters have come to realise 'the truth' about all this? Far better educated people seemed to be working their socks off pursuing the SM theories, the very ones whose opinions I then found myself seriously tempted to dismiss. Was there some more obvious explanation? Had I had a brain transplant and received a good one instead? … Had I simply flipped my lid? … Had I gone on walkabout in a 120˚F temperature? … Had I fallen down and hit my head? … No, none of these, it was all making real sense. It took a little more time for everything to find its place, but it did, so I buckled down to enjoy what I then could see was going to be one exceptionally interesting ride!

In terms of my own abilities, I was confident I could find what I needed from online research, science books, published papers and articles from respected electrical and plasma scientists and authors. I have to say that most of the people and their work I came across do support either one or many EU theories to significant degrees – they seem receptive to considering these and other theories. I saw this as being okay because I had previously put in just as much work, if not more, by looking at the work of supporters of the SM side of the debate. Through this process I was able to consolidate my thoughts enough to become convinced that my view of the two models was an acceptable one that should be taken seriously. Criticism might well be levelled at me as a lay person but I knew that my newly acquired understanding was backed by an impressive array of proven theory, solid evidence and the respected opinion of some very significant people indeed. Further to this, I had also found that 'best practice' existing as a natural process within the halls of yesterday's and today's astro-science research bodies, had not been and should not be, taken for granted as always being the case. Surprisingly, and sadly, it seems that things have not been that way for a very long time; since the early 1900s, in fact.

In terms of what I felt I could do personally about this, I decided that because I could see a need to tell others that the information was there to be had, and to provide it to them, I should summarise its major aspects into a simplified form and make that generally available. With this as an objective and the self-belief that I have some ability in producing less intimidating language from the sometimes difficult-to-understand concepts of others, I duly came up with a plan for this book. I considered it important that I use non-complex language and helpful graphics in describing the theories involved. Overall, the book would be my personal summary of how our universe actually functions and a cursory account of how in my opinion we have been led up the garden path by organisations, institutions and certain individuals, who for a very long time have become heavily invested in activities that have evolved with and supported the gravity-based SM.

## The Book

The story emerges from late in the 1800s and brings us to today as we leave behind the first decade of the 21st century. It begins with a summary of what we are currently told about the workings of our universe then goes on to look briefly at major controversial aspects of that story.

Described then is the background to how today's astro-science has evolved through what some would say are less desirable but understandable aspects of human nature. Thoughts are offered on how this may have come about and why the situation remains unchanged today.

The next piece goes into some detail on major questions around the theories and deductions currently supported by today's astro-science. These cover the well-known and thorny areas of black holes, dark matter, dark energy, neutron stars, quasars and more. I then go on to outline the major problems these aspects of theory now face through the ever increasing number of questions about them that the SM is unable to adequately answer.

I prepare then for the rest of the book through a review of some basic science and electrical engineering theories, this being a reminder for some which should help with getting to grips with the basics of the electrical and plasma theories described in the rest of the book.

I then expand on the background to all this by highlighting the significant and honourable 'other scientists' and their work, these being just some of the important characters associated with the electrical and plasma theories which underpin the EU model. Even today, the work of these and other people is often suppressed, ignored and occasionally attacked by the science establishment.

I then consider structures, objects and events in our universe from the EU model perspective and look in detail at the answers it offers that address the questions previously raised. We also explore the EU model as a more rational and logical basis for understanding what we observe and detect these days in space through our earth-based observatories and space-based satellite instrumentation.

The big picture that opens up is breathtaking and the serious questions that exist around the preoccupation that astro-science has with highly questionable theories will be seen as profound and puzzling in terms of why those questions still do not receive a fair and balanced public airing. I will attempt to give a sense throughout the book of the shared frustration that exists around this by posing the following questions ...

- Why has this situation come about and why are we justified in being suspicious of information coming from today's astro-science establishment?
- Why are the self-serving motivations of some being allowed to support a story about our universe that is not in line with observed reality?
- Why do we not honour the people who have devoted their lives in alternative ways to scientific progress and to what is obviously now there to be judged as more productive and meaningful research?
- Why do we continue to teach our children 'truths' that are actually only ideas and theories that have clearly been shown to be highly questionable and some even wrong?
- Why do we continue to bestow honours and funding on people who put time into complex speculative work based on bad theories which achieves nothing in terms of practical steps forward?
- Why do we never hear about the scientists and research workers who continue daily to work on 'science with substance' but who are forced to work in the shadows?
- Why do we fail to see through mathematical smoke screens and theoretical nonsense to appreciate the much simpler common sense explanations in front of our eyes?
- Why does the public not pay more attention to these things and what should we be doing about this?

Please understand as you carry on that I do not wish to come across as being too big for my boots, and neither do I intend offence to any reasonable person. I only want to tell a story that I have come to passionately believe in. Whatever anyone decides to do with this information is completely in their own hands. In writing what I have, there has been no desire to compete with those who are academically qualified in astro-science, theoretical mathematics, or indeed any other discipline. I understand the devotion of scientists to defend their dedication to the theories they choose to support. I am not able to argue against the complicated mathematics and detailed scientific 'evidence' they are no doubt capable of putting on the table. I am simply attempting to present in this book my own interpretation of the work done and conclusions reached so far by professional scientists and researchers who have obviously found substantial reason to put their names and reputations on the line by supporting the EU model.

In addition, my desire is to by-pass the blinkered and stubborn academic types, especially those that have their heads in their own dark places. These are the people we see preaching scientific opinion as if it was done and dusted fact on the television and through all forms of the media and on the Internet. They tell us with their words and through their style of speech that it is all either solved or close to being solved and infer that the SM is already accepted and correct. In my polite opinion, these people are misguided when they do this and I see them as also being disrespectful to people's intelligence. The problem is, we the public do not see that.

This is where I want to appeal to the common sense and logical thinking capabilities that the public at large have in great abundance, especially when faced with the opportunity to 'smell a rat!' We have all experienced situations where we know we are being spun a yarn but have felt we could not or should not say anything due to either a lack of detailed knowledge, a lack of confidence, or perhaps through some misplaced feeling of respect for those presenting the information. Well, we have one of these situations in front of us right now, so what do we do? If we say "*it does not affect me*" or "*why should I pay attention to this?*", then there are some serious points that might motivate us to think a little further. What about the quality of the knowledge we want to pass on to future generations? What about the need to force an accounting of the tremendous amounts of funding that pointless scientific research has received, and still receives? What if we just wanted to do the right thing because we know fundamentally that is what we should do? Perhaps the good people in positions of influence within science, those who could bring about real change, could think hard on these things.

Even within their own ranks, the voices of honourable scientists are there with their feelings of embarrassment and discomfort with what the science establishment, especially in theoretical astro-science, has been getting away with by doing and saying everything they can to perpetuate the story we have today. Tellingly, it is easy to explain the lack of clarity in the information that flows from astro-science these days, for they, themselves, are greatly confused. The astro-science establishment's bluff is being called, and we, the interested public, should stop giving automatic respect and acceptance to the patently wild ideas that some in their ranks present to the world. In my opinion, therefore, we need to consider that these people have at best been misled or at worst are self-serving and dishonourable.

I have made great efforts to peek behind the veil of mystery that surrounds today's theoretical astro-science, and to an extent I think I have achieved this. I now want to provide an opportunity for others who are on their own journeys of understanding in these matters to take something from what I have uncovered. As I have already said, but feel must be said again; we should do everything possible to ensure high and accurate standards in the information we provide to our children. We must get the facts that we teach them straight. Apart from doing a disservice to scientific advancement, anything less will set our children up to fail or be less than they could be in the future.

# 1 | What the problem is

I will begin by providing a reminder of the story currently given out by astro-science about how our universe came into being and how it works. Please note that this is not the story I now support, it is the one commonly promoted today by astro-scientists, authors, educators and elements of the media.

Most of us will have heard of the Big Bang [1-1], but what exactly was it? The name was invented in the 1900s during the time when ideas in astro-science about gravity and other things came together as the Standard Model (SM) of our universe. It was 13.7 billion (13,700,000,000) years ago that the Big Bang event is said to have started the creation of our universe. It took many millions of years for the sub-atomic particles produced as the only result of that 'explosion' to be drawn together to form atoms of gas. Then, with the assistance of gravity pulling that simple gas together, vast regions of it formed in concentrations that had shape and dimension.

What would the moment of a Big Bang look like?

Whatever graphic attempts are made to depict this event, not one of them could ever hope to produce a useful visual representation. This is because nothing we are familiar with would have existed then, not even light. 'a token image' © author

Gravity also brought together enormous gas 'stars' in incredibly large groups. This first cycle of giant star production ended with those stars 'dying' after rapidly burning their gas as fuel. As they expired, their matter collapsed under its own gravity then exploded in a process that fused their gas atoms into other atoms to form even heavier elements. As the force of gravity pulled those heavier elements and gases together, smaller, denser stars began to form. Unevenness in the distribution of these second generation stars allowed their combined gravity to gather those stars in vast separated collections until eventually, individual galaxies formed that contained billions upon billions of these second generation stars. At the same time, this process hoovered up vast amounts of left over gas and dust and included this in the structure of the formative galaxies. In all the time that has passed since then, and even now, the universe has continued to expand and the galaxy groups have continued to move away from each other. Inside the galaxies, stars of all sizes, types and compositions have constantly taken part in their own similar explosive cycle of birth and death that has given us the amazing structures of gas and dust we see today in space that we call Nebulae. This violent process is also responsible for creating the heavier elements needed to form planets, moons and all the other debris that now populates the space we know. The objects and events we detect with our telescopes and sensing equipment, plus those that are only theorised to exist, all have unique explanations within this story.

Each and every one of these things is explained to us through the gravity theories of the SM of the universe. I now end this story that is unacceptable to me but there are some other things to say before we move on to talk about the problems that many people have with it.

---

## The human view

It is important to appreciate how we humans tend to think. For those of us who like to think deeply, our truly limited abilities only become apparent when we attempt to consider what can be seen out there in space. We ask ourselves if we will ever be able to appreciate the fact that what we are looking at is only a tiny snapshot in time in the overall existence of our universe, whatever that scale of time may be. It is sobering to think that the best impression we can achieve would have no great relevance within a single tick of the universal clock; it would be no time at all. However, as we humans naïvely tend to do, we look out there with our current level of understanding and equipment and feel that it all does or can make sense to us. Gathering knowledge is undoubtedly among the best of pursuits, but for some of us, our know-it-all tendency goes further and leads us to think we actually do understand it all. In truth, we might actually be incapable of considering the unimaginably vast amount of time and number and types of events that have gone before us. It is during that period gone by that everything we now observe in space was formed and moved around over unimaginable distances under the influence of unimaginable forces that were acting on unimaginable amounts of matter travelling at unimaginable speeds. I think that we are very arrogant indeed if we believe we have a good grasp of what our universe is about and what it has gone through during countless eons. Is it not typical for us humans to be cocky about our capabilities and levels of achievement?

We have been told by astro-science that the things I have mentioned: the gas, the first giant gas stars, the second more solid matter-inclusive stars, the dust and the galaxies that subsequently formed, all have been influenced by one force alone; the force of gravity. This view has lingered with us for many decades and has not changed in any significant way. The seeds from which it grew were sown back in the 1700s when an understanding of the effects of the force of gravity was first developed. Scientists and astronomers welcomed this for they saw the beginnings of a 'fundamental golden theory' that they could take forward and use to develop their own lines of research. This focus was indeed nurtured by many through the rest of the 1700s, 1800s and into the early part of the 1900s. And the theories since derived have turned out to be the pillars on which a questionable story of the origin of our universe has evolved to become, superficially, a very believable one. Strangely, it seems to us today in the 21st century that this 'trustworthy information' has always been there, for most of us have just accepted it without question as the story of how the universe came to be and how it currently functions.

Knowledge of the force of gravity was indeed very important to science and it was at the heart of many ideas that evolved throughout the 1900s. Through prior auto-acceptance of the major role that gravity played, other ideas associated with it were also easily accepted as scientific fact. We will look at some of these other theories that involved gravity and which evolved as part of the SM during the 1900s.

Here we have Dark Matter, Dark Energy, Inflation, White Dwarfs, Pulsars, Neutron Stars, Magnetars, Black Holes and Nova events. There are more, but these will do for now. First I will tell you that we do not know what any of these things are supposed to look like, so the following graphics in no way represent actual physical objects or events. They are included only to help in imagining the concepts being described.

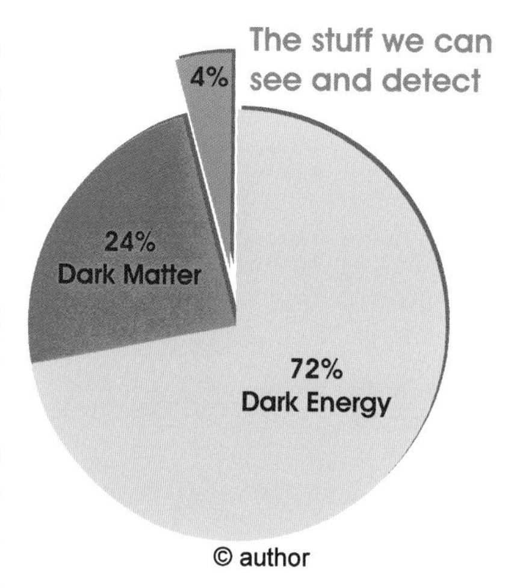


© author

Before talking about Dark Matter and Dark Energy, I need to say something about the composition of our universe. This is the mixture of 'stuff' and 'forces' that astro-science says can explain its appearance and behaviour.

The amount of 'physical matter' in the total universe that can actually be seen or otherwise detected or estimated by us, has been calculated to be only 4% of the total amount of matter that is actually present [1-2A]. This estimate is based only on objects and events that have resulted from gas and dust and the effects of gravity acting on those forms of matter. What then makes up the other 96% that mathematics tells today's astro-scientists is there? This is where their calculations have produced a division of this figure to say that dark matter is around 24% and dark energy is around 72%.

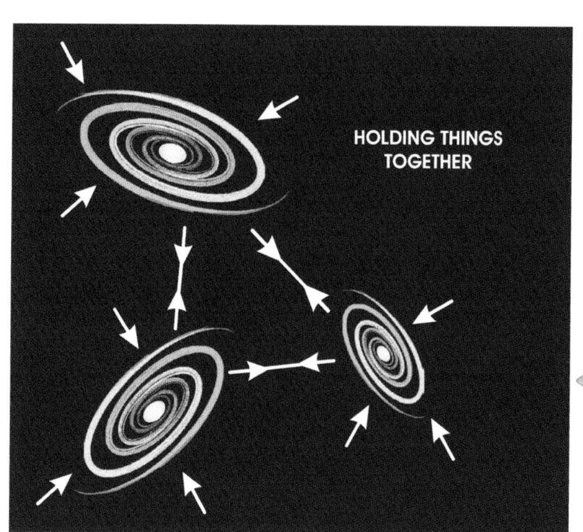


**Dark Matter** is a name chosen by the Bulgarian astronomer Fritz Zwicky. He came up with it to describe a special type of matter he thought was needed to explain the binding force that seems to exist within and between galaxies. For Zwicky, this would be the force that keeps structures such as galaxies together and the one that in an overall sense stops everything from flying apart; this 'flying apart' aspect being the problem Zwicky wanted to solve in 1933 because it was a problem for the SM at the time.

◀ Galaxies and the Stars inside being held together © author

Zwicky calculated that if dark matter was placed at strategic locations within and around galaxies, the gravity it produced would save the original ideas on galaxy creation and behaviour.

He believed that by including dark matter he could account for the missing gravity that seemed to be holding things together and allowing everything to appear as it does. This extra gravity would be in addition to that already produced by the existing visible matter of galaxies and the 'super-massive Black Holes' that were imagined to exist at their centres. So, when all the gravity was added up the result was made to suit the requirements of the mathematician's calculations and so everyone was happy. Zwicky had come up with an explanation for the missing amount of gravity, one that was very acceptable to the astro-science community of the day because it explained the things they had come to believe. Interestingly, however, Zwicky's dark matter just happened to be invisible and totally undetectable. So, in the eyes of those few on the outside who saw this as a problem, the idea of dark matter appeared to be just a 'convenient fix'.

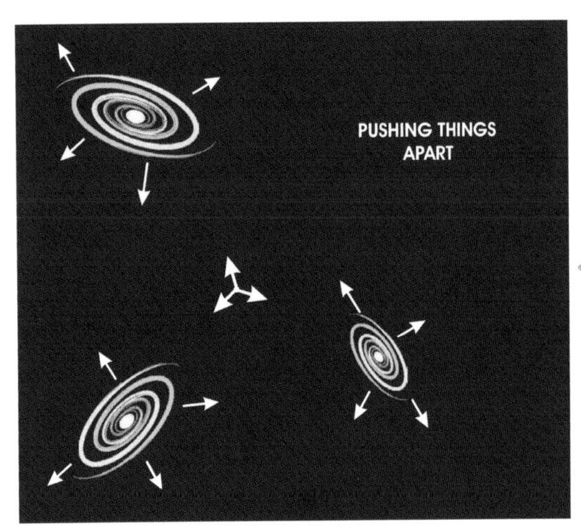


**Dark Energy** is a name coined in 1998 by the American cosmologist Michael Turner. This invention was required after the introduction of dark matter in order to counter a problem that had arisen from further observations and a subsequent period of great embarrassment.

◀ Galaxies and the Stars inside being pushed apart © author

The problem was that galaxies, while keeping their form as individual structures, had been deduced to be moving away from one another at an accelerating rate, so the universe must be expanding! How could a force be accounted for that would cause this; what could make this happen if there already was a binding force to hold things together? Turner decided that a new 'anti-gravity' force was required. This force, when dominant, would act to push bodies away from one another and therefore accelerating expanding space could be explained. His solution for this was the idea of invisible dark energy! This was a 'great discovery' for it fitted very well with Zwicky's idea of dark matter. So together, these two 'dark entities' were able to make up the invisible 96% of matter that previously could not be accounted for. No trace of any of this dark stuff has ever been found, even though our best equipment and lots of scientists have been funded to look for it over many decades. It is abundantly clear now that both varieties of dark stuff were invented purely to save the gravity model of the universe [1-2B].

The associated idea of **Inflation** [1-3] is also involved here. This was proposed by the American theoretical physicist Alan Guth in 1980 to address an important issue that had arisen from the fact that, no matter in what direction we look out into space, the level of background energy coming towards us from the furthest places was being measured at very similar levels. A Big Bang type explosion could not have made things so smooth in energy terms on such a vast scale, so this again was a serious problem to explain. Guth also said that in order to account for how the furthest observable points in opposite directions in space could have originated from the same location 13.7 billion years ago, a super-luminal (faster than light) event must have taken place just after the instant of the Big Bang. Inflation was Guth's imaginative invention for this purpose as it could explain the super smoothness of the background to deep space. Once again we find a single idea swiftly accepted as another essential element of the SM. In reality, however, it was just another 'sticking plaster' to prop that troubled model up.

**Quasars,** [1-4] first discovered in the 1950s, are the mysterious 'firecrackers' of the universe. They 'appear' to us small, bright and energetic and are assumed to be young formative galaxies.

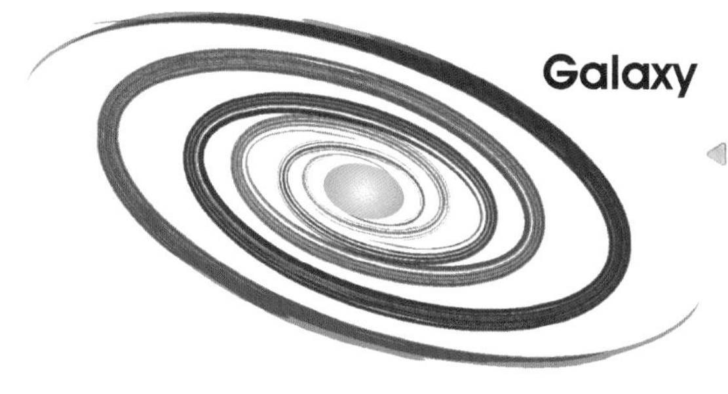


Young Quasar and Old Galaxy © author

Many of the hundreds of thousands of quasars discovered so far are described as being far away at the limits of our observational capabilities. Due to the calculated amount of time their apparent bright light has taken to get to us, their visual appearance has been judged as being how they appeared when the universe was very young. These quasars (Quasi Stellar Objects or QSOs) are another puzzle for today's astronomers to solve. This is mainly due to evidence accumulated by astronomers who are not part of the mainstream, but also some who have been, which suggests that quasars are not very far away at all and that their origin and appearance are explicable through theories unrelated to gravity. I will say more in detail about quasars later.

**White Dwarf Stars** [1-5] are said to be dense Earth-sized remnants of dead stars, which in simple terms weigh around the same as our Sun. (We really should be talking about the 'mass' of objects instead of their 'weight'.)

A White Dwarf Star? © author

Many of the white dwarfs that are near to us appear not to be especially bright. Their extreme mass (density) is said to result from the action of the force of gravity collapsing their original star's gas so much that the resulting core is compressed to an extent where it ends up having the appearance of a small whitish glowing ball. These stars are typically described as being the 'core remnant' of a certain size of star that has gone through this process and thrown its outer materials off into space.

They are further thought of as stars in the process of cooling because no ongoing energy generation process exists within them to maintain their temperature. The eventual fate of a white dwarf is for it to become an undetectable cold and dark cinder (or Brown or Black Dwarf, as some call them.) One puzzle with these objects is that they are often found in the company of two or more companion normal stars, where each has been assessed through standard procedures as being very different in age from the white dwarf. Our own relatively small Sun is a star in the category said to be destined to eventually produce a white dwarf star.

**Neutron Stars** [1-6] allegedly result from the death of a much larger variety of star that also collapses inward under its own gravitational force. Their final diameter is estimated to be in the order of 20 kilometres and their mass to be around one and half times that of our Sun.

◀ A Neutron Star? © author

The atoms of this star's matter are squeezed together to such an extent that their own structure breaks down to leave nothing but neutrons. Neutron stars are therefore said to be completely made of neutrons; these being the heaviest (most massive) of the sub-atomic particles. It is speculated that a single teaspoonful of neutron star material (so-called 'Neutronium') would weigh billions of tons here on Earth. These bodies are apparently detected in a couple of ways. One method uses the gravitational influence they are judged to exert on matter and objects in their immediate environment. The second method is based on the assumption that these stars are the source of electromagnetic pulses; this being the most common indication that gives their presence away. It is this 'pulsing behaviour' that has also given some them the label of '**Pulsar**' [1-6]. The high-energy radiation that Pulsars emit on an extremely regular basis is an action often described as being similar to that of a rotating 'Lighthouse Beam'.

A Pulsar - a rotating beacon? © author ▶

It is further supposed that this beaming action actually consists of two beams of narrowly directed radiation that sweep individual paths through space. Think about the rotating light on an emergency vehicle and you will have a fair idea of this effect. On occasions through some as yet unexplained mechanism, Pulsars seem to be endowed with an unnatural rate of rotation. I have seen in some articles this speed being quoted as being as high as 72,000 rpm. This type of claim should make a normal person sit up and wonder about how a body kilometres in diameter could possibly rotate at such a speed and still remain in one piece!

Neutron Stars and their Pulsar cousins are considered to be common in space. However, we must keep in mind that they have only arisen from a gravity-based interpretation of observations backed up by maths which conveniently 'proves' they are real. None of these objects have ever been observed as actual bodies in space so the above 'guesswork', for that is what it must be seen as, is all that stands as evidence for their existence.

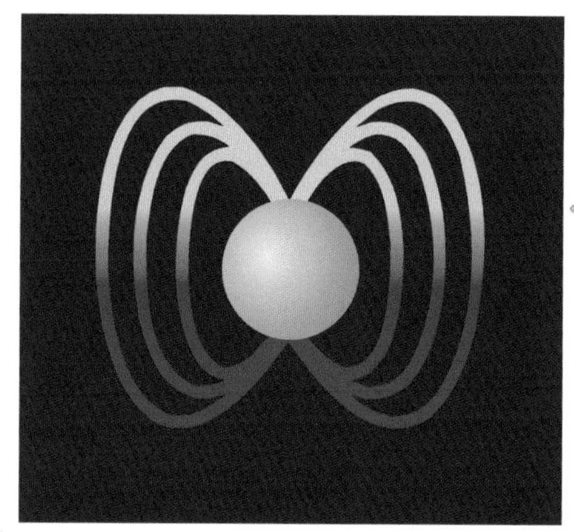

A Magnetar? © author

**Magnetars** [1-6] are a relatively new type of object for astronomers to play with. They are supposedly formed through the same process of star material collapsing when the gas fuel of the star runs out.

The detection of these supposed bodies is again through the effect they seem to have on objects and matter in their environment, and especially in this case through the high-energy X-ray and gamma ray radiation emissions that are said to come from them. Millions of Magnetars are claimed to exist in our Milky Way galaxy alone, with lifetimes estimated to be around 10,000 years. Please remember, however, all this information is theoretical; it is based purely on observations that today's astro-scientists have chosen to interpret in an *ad hoc* way to match observations that support the SM.

Now we consider **Black Holes** [1-7]. These were thought up by Dr John A Wheeler, emeritus professor of Physics at Princeton University in the US. He came up with the theory but not the name as an explanation for why certain galaxies are observed to have very high-energy radiation coming from their central cores and how these galaxies were able to hold themselves together.

Once again we have an object that arises from the collapse of a star; this time, very large stars indeed, or more than just one star. The theory behind them was an exciting one so it grew rapidly to include the idea of 'super-massive Black Holes'. These horrendously powerful gravity monsters are said to be made up of millions of collapsed stars, all having come together as one unimaginably enormous Black Hole. This idea fitted snugly to help explain the great amount of gravity, in addition to that supplied by dark matter, that would be required to perform the substantial job of holding a galaxy together. However, we must note that no Black Hole has ever been observed and no experiments on Earth or in space can currently be carried out to prove they are real. As you will see, Black Holes are another product of wishful thinking encouraged by gymnastic, unrestrained mathematics.

A Black Hole? © author

White Dwarfs, Neutron Stars, Pulsars, Magnetars and Black Holes are all said to have the same origin, and coming across this variety of names can be understandably unhelpful and confusing. At the low end of the star death scale we are told small stars produce white dwarfs and at the high end whopping great big stars produce Black Holes. Then we are told that all the other objects mentioned here take their place between these limits. So, what more is there to say about star death? This brings us to the terms **Nova** and a **Supernova** [1-8].

As the names imply, a **Nova** is seen as a smaller version of a **Supernova**, where both terms are taken to represent events, not objects. A nova event can confusingly be found associated with both the birth and death of stars, each of these being said to throw material off into space in an energetic brilliant flash. The energy involved is reckoned to be stupendous, hence the apparent brightness and high levels of radiation emitted. The vast dense region of dust and gas that results, expands rapidly with a shock-front at its outer edge. This high-energy wave front moves powerfully away from the central event to collide with all other surrounding material. When this happens, the matter struck by the shock-front swiftly becomes heated (energised) and is forced to glow to such an extent that it gives off high levels of radiation. Through this process vast regions of glowing gas and dust are formed which we now call **Nebulae**. We must however remember this is a gravity-related explanation. It relies on mechanical shock alone to produce great amounts of ultra-violet and X-ray radiation, an idea that has in fact been shown as inadequate in terms of the energy levels actually required for such processes.

◀ A Nova event? © author

A Supernova event? © author ▶

Larger Supernova events are said to result from the more powerful 'death events' of large stars, a process through which fantastic amounts of radiation are again claimed to be emitted. In addition to these gravity mysteries, astro-science has adopted other theories to support the SM. We will be looking at some of these shortly but first, we will pause to consider aspects of the public image of astro-science that has evolved in combination with these theories. This image now strongly implies that the work of today's astronomers and astro-scientists is not accurately on-track, so why might that be?

The fact is that a significant number of 'theories' are only unproven hypotheses (good guesses, if you like) that have not achieved the status of being proper theories. And some of these, even though provably dubious, remain accepted as integral components of taxpayer-funded projects. Furthermore, a list of serious questions as points of objection has built up around these 'theories' from outside the astro-science mainstream. This list has been in place for many decades and has not by any stretch of the imagination been addressed adequately.

Other theories associated with the SM are also presented with great confidence and I will be going into these in more depth later, but for now, a few of them are:

- The thermonuclear model of the Sun
- The way that 'redshifted light' has been used to calculate speed and distance in space
- Comet composition and behaviour
- Cratering and surface scarring on planets, moons, comets and asteroids.

**The thermonuclear model of the Sun** [1-9]. The Sun and all stars are portrayed as being some form of gigantic thermonuclear fusion reactors - a type of ongoing self-controlling nuclear explosion.

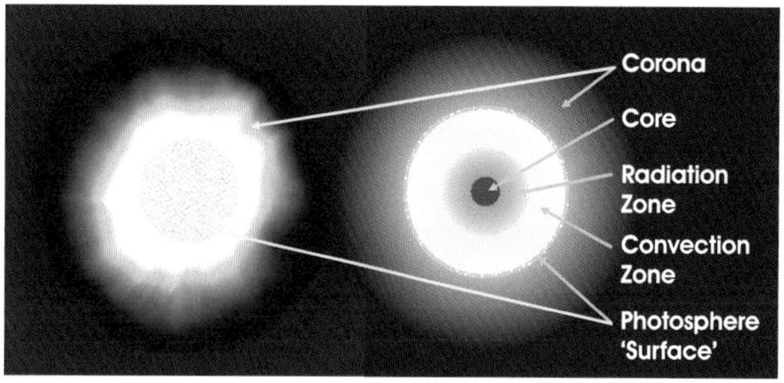

The 'layers' associated with the Sun © author

This powerful nuclear reaction is said to produce enough high-energy radiation for the matter around the core of the Sun to be forced outward, thereby achieving a level of mechanical balance against the inward pull of the Sun's own gravity. Because this effect is the case in every direction, the Sun (and all stars) are able to maintain a spherical structure. As a theory, this seems to be quite straightforward but because of real-life observations, there are some crippling questions that supporters of this model have yet to answer; questions such as:

- What explains areas of the Sun being relatively cool where they should logically be hot?
- How do magnetic fields within the Sun and those associated with visible sunspots specifically, shield from us the heat generated inside?
- If the Sun really does have a continuous nuclear explosion going on inside it, why then do we see only a small level of a particular 'nuclear action tell-tale' type of radiation being emitted where this should actually be at a very much higher level?

**Redshift** [1-10], in the context of astronomy, is the term applied to light as it is seen coming towards us from objects that are themselves judged to be moving away from us.

How redshift and blueshift come about in terms of the stars we observe © author

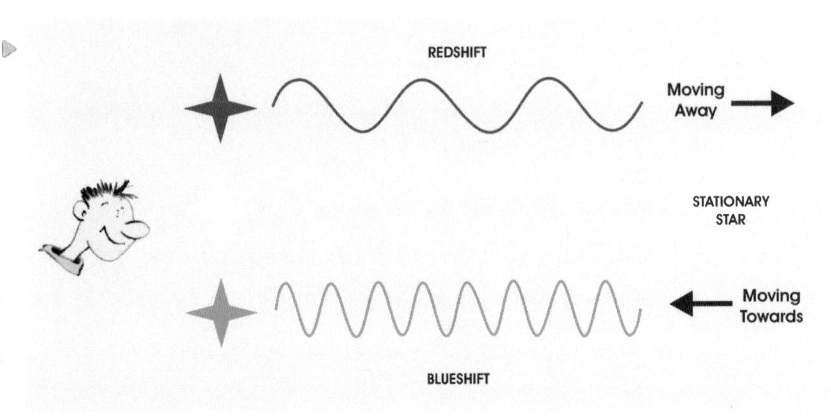


In the case shown here in the graphic we see three things. At the top we have a star moving away from us. In the middle, the same star is stationary, and at the bottom, it is moving towards us. In each case the natural light from the star will take on a particular appearance when we observe it.

Where the star is moving away, its light wave is stretched out and so is said to be 'shifted' towards the red end of the visible spectrum; hence the term, redshift. Where the star is not moving, its light wave has not been affected so the light is natural or 'unshifted'. Where the star is moving towards us, the light wave is compressed or 'shifted' towards the blue end of the spectrum, an effect that we call 'blueshift'.

In the 1920s the famous astronomer Edwin Hubble found that looking beyond our Milky Way galaxy it was not the only one in the universe and that by comparing the size and apparent brightness of the other galaxies he found, they appeared to be moving away at an increasing rate. He deduced this relative acceleration through the application of his 'redshift theory'; the one described above. However, as it turned out, Hubble himself eventually doubted the reliability of this theory due to further aspects of his work that gave him reason to question what redshift actually did represent. He even went as far as talking about these doubts with colleagues, but it was too late; redshift, together with Hubble's respected name and formidable reputation, had already spread throughout the mainstream as a fundamental and accepted 'reliable tool' for use in astronomical calculations.

In the astro-science community redshift had come to be seen as proof of three things: One, that an object's speed of travel (recession) away from Earth could be established; two, that the speed of these objects was apparently increasing as their distance away from us increased, and three, that when 1 and 2 are taken together, the universe itself could be interpreted as expanding at an increasing rate [1-11]. One further and very profound assumption was simplistically taken from what redshift was assumed to represent. This was that, since everything was apparently travelling away from everything else, then all the matter in our universe must have been located at the same central point in the very distant past.

The Big Bang - an impossible explosion in nothingness? © author

This superficially logical assumption was later to become the acorn that grew into the notion of 'The Big Bang'. This name was coined by the famous English astronomer Sir Fred Hoyle during a radio interview as a suspected term of derision due to his rejection of the idea. After all, and amongst other things, the idea was that everything in our universe began with its origin at one point within ... nothingness! However, if the ideas around redshift were correct then there was indeed room to assume that a Big Bang could have happened and that by implication for so many good folks, this would also suggest that the Bible's moment of creation had been proved by science, no less! Understandably, religions all over the world went cock-a-hoop at the idea, but as we will see, redshift has been under critical inspection for a long time. Many more and far better quality observations and analyses of data have, over the years, shown that the redshift factor and what it has been used for has resulted in ideas and practices founded on very shaky ground indeed. Take a moment to think how serious a suggestion this is. It brings into question the whole idea of a Big Bang event and religion's reliance upon a supposed scientifically proven moment of creation.

The story of **Comets** [1-12] is a great one to further highlight the intractable problems faced by today's SM. Comets have long been described in the following terms: They are lumps of loosely-packed, dirty, icy material that travel on long elliptical orbits within our Solar System. As they get close to the Sun they begin to sublimate (i.e. they convert their solid icy surface material directly into gas) as part of a process that accounts for the enormous 'glowing cloud' often seen surrounding them and the long tails they produce behind them. All the material expelled from a comet's surface that is showered into space as gas and dusty debris is said to occur as a result of this process.

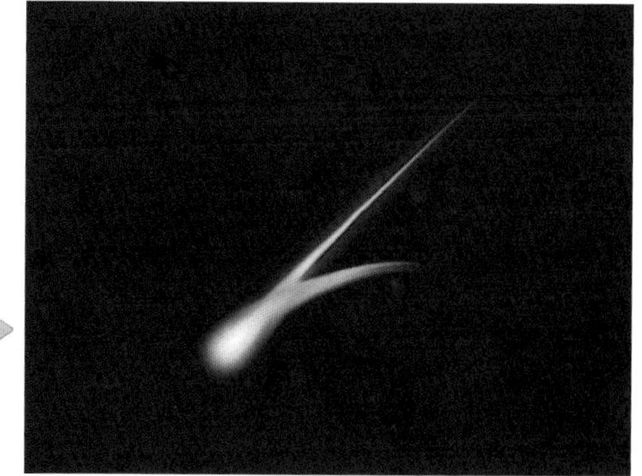

The 'tails' of a comet © author

Over recent decades there have been many close-up observations of comets and a much better analysis of their behaviour than had previously been possible. The questions that have arisen from the results obtained now stand as serious points against the dirty snowball model. The evidence now is that comets are far different in their make-up than we have been led to believe. In view of this better quality information being available, it seems incredible that the astro-science establishment continues to hold on to its dirty old snowball theory. Again, we will be looking in detail later at the evidence that explains what comets really are and why they behave as they do.

We have all seen **Craters** [1-13] in pictures of the Moon's surface. There are craters on Earth as well and there are craters on almost every other solid body in our solar system – including asteroids and comets.

Themis Crater on Mars - Image courtesy NASA/JPL-Caltech/ASU

The explanation for how craters were formed is that long ago, swarms of flying rocks bombarded the surfaces of larger solid objects at great speed and with enough force to cause the circular pockmarks we now see. Once more we have gravity at the root of this story as the attracting force between the large body and the smaller projectile. This explanation seems to fit rather well but I would suggest that most of us have just accepted it as believable without thinking too much about it. With the subject of craters, we once again have a situation where new evidence brings an old, simplistic idea into question.

The apparent truth behind cratering is fascinating and has a fundamental link with EU theory. We will see later that the events most likely to have brought cratering about are not at all what folks might suppose.

The process of **Surface Scarring** [1-14] on planets, moons and other bodies has also been attributed to accepted theory. Most of us will have heard about the Earth itself having a molten core and about the movement of its surface tectonic plates. This tectonic movement is what is said to have given us the continents with their mountain ranges, deep valley gashes and other physical features - all of these being produced as the plates either slid past or crashed into one another and where areas of some of them were stretched apart. It so happens that this earth-centric type of thinking has been simplistically carried over to come up with explanations for similar features observed on the surfaces of other planets and moons, etc. The worrying thing here is that this 'short-cut' approach has produced explanations which yet again have been readily accepted by most people without question because they are seen as good ideas that make sense.

The presumption of surface crust movement as the root cause for mountain ranges, tears, gashes and other blemishes on every solar system object is a bold step, and one that has rightly been brought into question. It has been shown as impossible for many solar system bodies to ever have had a geological mechanism that could account for surface movement in the first place – so how then could the mountains, valleys and various gashes we observe on their surfaces have come about? The aspect of surface gashes on many solar system bodies also shows up through what are commonly referred to as rilles.

Rilles are winding channels, seemingly gouged into the surface of solid bodies such as the Moon, Mars and Mercury. The most common theories explaining these are that either water must have existed on these bodies in the past and that the flow of that water has been responsible or that those planets and moons have somehow been honeycombed by sub-surface lava flows which in turn form long tunnel-like voids close to the surface, the roofs of which collapsed over time to produce the rilles we now see.

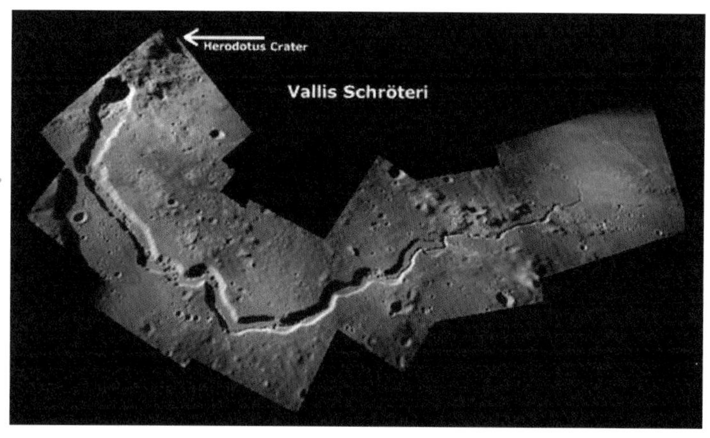


Vallis Schroteri on Mars - Public Domain US Gov

If these types of event really occurred, then explanations should be provided as to why some of these rilles have no apparent entrance or exit points for water, and why there is no typical evidence whatsoever that indicates surface collapse.

In addition … Why do rilles often turn at sharp 90 degree angles? … Why do none of them indicate the same patterns left by how water flows here on Earth? … Why are many of them seen to be crossing each other, and why do some of them indicate that if water was involved, then it would have needed to flow uphill? Again, we have allowed creative astro-science to drive the bus, a situation that has produced answers that just do not stack up. More plausible explanations exist for what has occurred to cause these craters, gashes, gullies, rilles, and other surface features, and we will be looking at these later.

The issues pointed out in this chapter are not the only reasons to be dissatisfied with the SM, or indeed, with other things we have been told by the astro-science establishment. It is surprising that despite the array of far more credible theories, the astro-science community continues to energetically promote highly questionable theories. This of course affects science itself, but we should also remember that it affects our schools, colleges and universities. Misinformation pervades popular books and articles promoted as educationally sound, and it is found at the core of science support materials and in information presented to the public through television documentaries and through a multitude of other information conduits. One would think therefore that this apparent educational and developmental misdeed is something that deserves to be challenged.

Having identified a few of the problems with standard thinking about our universe, could it be that I am just re-playing some tune picked up from a few old scientists who have been left feeling dissatisfied with how astro-science has developed for them personally? Alternatively, is there another reason just as weak and pointless as that? No, I can assure you that the questions being raised here are all well known and very serious indeed.

Despite this being the case it seems that no official support can be expected for open discussion of these points, any one of which would be negatively disruptive for today's astro-science establishment. The issues raised are especially ignored because if the public were to pay more attention and seriously support them, then an unwelcome spotlight would be cast upon the theories that allow the SM to exist. Moreover, and to add to the woes of astro-science, we are now inundated with good data from much improved earth-based and space-based sensors which in many cases provokes even more detailed, awkward questions for astro-science to answer.

To round off this chapter, here are a few more things from our observations of space to consider:

- The puzzling clues now provided to us through greatly enhanced images of ultra-violet, X-ray and gamma ray emissions coming from our Sun and from its close environment. The analysis of these emissions seems to stand as evidence of the classic behaviour of electric currents and their associated magnetic fields having a major presence and influence on and around the Sun's surface. [1-15]
- The detection of much greater amounts of high-energy X-ray and gamma ray radiation from deep space than standard gravity theory predicts or is able to account for. [1-16] [1-17]
- The fact that the space probes Pioneers 10 and 11, launched decades ago, are now leaving the solar system and have been found to be a quarter of a million miles off course, slowing down and meeting the unexpected behaviour of charged particles. What force could be affecting them so far away? [1-18]
- The confusion of hot and icy materials on the surface of Saturn's moon Enceladus. What could possibly cause this? [1-19A] [1-19B]
- What appears to be a hexagon-shaped aurora formation at the North Pole of Saturn, first observed by the space probe Voyager 1 and then confirmed by the Cassini probe in 2006. What force do we know of that has the ability to organise matter in a regular geometric shape like this? [1-20]
- And then back here on Earth we have the rather expensive white elephant in Switzerland. This is the multi-billion dollar LHC (Large Hadron Collider) that conducts experiments designed to identify the sub-atomic particle labelled the Higgs Boson that is said to be responsible for a dark force that remains invisible and undetectable to us but which is claimed to explain mass. The LHC project has absorbed a crippling amount of money that could otherwise have been used in constructive areas of scientific research or to comfortably run a small country. It is an impressive but ill-fated endeavour because again, it seems to be a product of blinkered science based on faulty theories. Nevertheless, it continues to enjoy unwavering support. [1-21]

Times are changing, and I look forward with optimism to the better use of resources for research purposes and to knowing answers to questions arising from that research.

# 2 | The thinking that got us here

I read years ago of how the 15th-16th century Polish astronomer and mathematician Nicolaus Copernicus came up with the idea that the Earth was not at the centre of the observed universe as people had previously been told it was by religious teachings.

*Nicolaus Copernicus*
*1473 - 1543*

The unfortunate implication of this was that, if we humans were not at the centre of things, then we would be nothing special. Copernicus saw this as quite a radical thing to suggest, especially as it went against the teachings of the Catholic Church, the organisation that had promoted the human race to a central role through its particular view of the universe. Therefore, being in fear of ridicule and the power of the church, he did not let this idea go public at the time he came up with it. The Catholic Church certainly did wield an awful lot of power and influence in those days and it would only be the plain stupid or the insanely dedicated that would actually make known any religiously contentious beliefs they had. If there was even a hint of opposition to the church's doctrine through what people said, bad things could happen to them, so it was only at the time of Copernicus' death that his work was actually published.

The Copernican Solar System © author

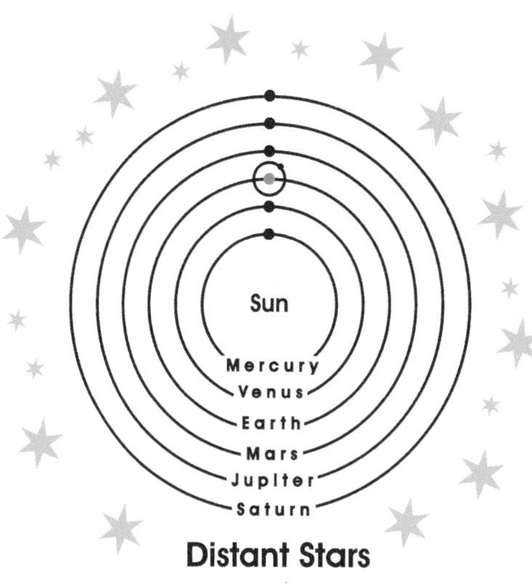


The Copernican model of the universe was that the Earth and the other planets known at the time, all revolved around the Sun in concentric circular (assumed) orbits. This model of our Solar System is the one we are all familiar with today. However, Copernicus got one element of his model wrong; he said the Sun was at the centre of the universe! This may have been because, already having determined the Earth's relationship to the Sun, he could see for himself in the night sky that there appeared to be a globe of stars of relative equal density surrounding us, no matter in which direction he looked. Considering the importance of those events, we could easily view them as the beginnings of sensible astronomy. What Copernicus had deduced about our solar system had at least been based on practical observations, basic science, and some common sense.

There have of course been very important contributions made to astronomy by other significant people such as Johannes Kepler, Galileo Galilei and many others before their time and since. These contributions, significant as they were and in most cases still remain, are not critical to understanding the main thread of the story I am telling here, so forgive me for limiting this foray into the history of astronomy to only what I do touch on.

◀ Sir Issac Newton 1642 - 1727

Things moved on in all aspects of astronomy until late in the 1600s the English physicist, mathematician and astronomer Sir Isaac Newton, came along with his theory of gravity. This described the rules by which all things on Earth tended to fall towards its centre and other situations where the action of gravitational attraction between two bodies needed to be taken into account. By the way, the event with the apple was an invented story - it never happened! What people usually do not appreciate about Newton's law of gravity is that it only describes 'the effects of gravity' and not 'what gravity is' or 'how gravity works'. It also takes no account of time, so it is considered to be an instantaneous force that is in effect everywhere at once. It is accepted now that Newton's law of gravitational force is fine to use at our solar system scale but perhaps not when it comes to very big galactic scales. This suggests that Newton's laws, testable in many simple ways here on Earth, are only part of a much bigger story. There was no reason to question Newton's ideas so they soon turned into theories and then into accepted universal fact. Newton's laws continue to be used in a qualified way, now that we understand them better.

Albert Einstein 1879 - 1955 ▶

The interpretation of Newton's laws which said they could explain all things was the case when young Albert Einstein came on the scene before the turn of the last century. He took the laws of gravity and expanded on them in ways that were eventually to become the hallowed gravity and relativity foundations of the current story of our universe. It is relevant to point out that, as Einstein himself had done with other folk's ideas, his own original work was added to and 'refined' by other scientists in physics and astronomy. In this respect, he was not completely happy with what some of his work eventually came to represent, so Einstein has in some respects been misrepresented. Even now his name is still casually associated with aspects of science that he considered to be questionable. Nevertheless, the man remains one of the heroes of science.

I have a great deal of respect for Copernicus, Newton and Einstein as the honourable and dedicated characters they seem to have been. There are other contributors to science who deserve a mention, but that would be something for a different book, not this one. If you are particularly interested to know more about these and other pioneers of science, then you should look into the ancient and modern history of astronomy, physics, cosmology and mathematics. There you will find many in-depth historical and biographical accounts that do an amazingly detailed and interesting job of describing the work of these and other significant people. Suffice it to say that the foundations for what we have today as the story of our universe were laid from the work of Newton and Einstein.

It was in 1905 that Einstein produced his first major theory that was eventually to become famously known as his 'Special Theory of Relativity'. This title came about in 1919 after a particularly supportive observation was carried out by one of his colleagues, Sir Arthur Eddington, that apparently verified what Einstein's theory claimed. In the theory, Einstein stated that among other things, the speed of light was constant at 300,000 kilometres per second (in a vacuum) for all observers, whether they were moving or not. What he really had in mind here was the assumed empty vacuum of space. Derived from this was the inference that the speed of light is the fastest that anything can travel at in the universe - a universal speed limit, if you like. Although debated at the time, Einstein's idea that light could travel through 'nothing' prevailed, so that which had become his Special Theory of Relativity stood strong among his peers and rapidly became accepted as scientific fact.

Later in 1915, Einstein came up with his second major theory, the 'General Theory of Relativity'. This was his crowning glory, in which he made a stab at laying out a geometric explanation of gravity. In doing this, he also came up with the assertion that our universe existed within a framework of what he labelled 'space-time'. The invention of this term was his way of describing our universe as an environment wherein time and space are related, or to put it another way, if you were able to change space, then you would change time, and vice-versa. These ideas seemed a bit crazy to some but made sense to others when the effects suggested were considered possible only when travelling at or close to the speed of light. Again, nobody seemed to have cause or motivation to object to Einstein's latest theory, so it stood strong for a while before becoming acceptable scientific 'fact'. None of this could be proved but it seemed that one plausible theory could stand with another, so the overall picture that developed took on a life of its own.

Note here that Einstein's theories really did spring from his own imagination. They had started as ideas from his personal 'thought experiments', as he called them; periods of deep thought that were also fuelled by aspects of work done by others. No fundamental experiments could be done at the time to prove any of his theories and it remains a fact today that in terms of what they claim on the grandest scales, they have still not been proven. This means that anything currently promoted as solid fact regarding the assumed powerful influence of the force of gravity on large scales can summarily be explained as only being the output of athletic, imaginative thinking. This fact alone should be enough to set some alarm bells ringing for us!

Having reached this point with gravity at the heart of the story, it is clear that whatever work done throughout the last century that has relied on these theories has in fact been conducted on the basis of one person's imaginative ideas, but with no real physical understanding of what gravity is or how it works. This, amazingly, has been the situation for more than 100 years! Ah, but Einstein was a brilliant scientist, I hear you say! That assertion is certainly not in dispute, but the fact remains that over the many decades since the first half of the last century, his ideas have been developed and moulded to create and support a very questionable story about the origins and operation of our universe. It is this story that certainly is being brought into question. I will point out here what in my opinion scientific research really should consist of …

What is scientific research? Well, people talk about following the 'scientific method' as a process for establishing solid evidence that can be taken forward and further developed. It seems, however, that at the sharp end of obtaining results from experimentation, personal opinions become excessively involved when the scientific method is discussed. The fact is, there exist general guidelines that scientists and researchers apply their own standards to in their efforts to obtain quality results. This tendency towards 'freedom to focus' highlights that in general, scientists and researchers can make up their own rules. However, for this same reason, they must rise or fall on the results they produce, especially if they are willing to have their work assessed by all and not just their close colleagues and sympathetic friends. Interestingly, it seems from what I have come across in my studies that much of astro-science's research in particular is carried out in opinionated minefields that can also be highly self-serving and protective while often being fiercely political.

If we take a step back and think of the hypothetical situation where only one particular theory; one line of thinking, was in place, how then could we expect progress to be made in our overall understanding if nothing was available with which to compare that single theory against? A thinking person would, I suggest, say that no progress would be made! This means that in order to progress appropriately, we must first have the conditions in place to formulate new ideas and to generate questions from these that probe other ideas. We need to have the chance to compare our work and theories with other work in order to identify relevant points of issue and become aware of alternative approaches. Note here how this differs from just dreaming up ideas in one's own mind under one's own rules with no checks or balances whatsoever. In general, the practice of entertaining without necessarily accepting alternate theories is what real scientists try to do in pursuit of useful knowledge. How then might they go about good research? Here from my own position is what I think it is reasonable to expect scientific research can be seen as …

Ongoing rigorous experimentation in the pursuit of new discoveries within and around the boundaries of sensible research that support the development and substantiation of acceptable theories and real-life observations, thereby showing the way ahead.

In order for science to progress healthily, it should be open to alternative ideas and constructive challenges. Emerging theories should be listened to and tested to their limits by examination. This should be done through consideration of the views of peers from within the discipline involved and possibly the views of experts from other disciplines, where appropriate. And as has been said many times by good scientists, there should always be opportunity for alternative theories to progress in parallel with, or to replace completely, an original theory.

It seems however, that the less rigorous and less professional members of the science establishment (in theoretical astro-science, cosmology and mathematics especially) have paid questionable heed to these things. Many of them seem instead, for their own reasons, to want to continue following what have been shown to be 'fairy stories'. In the words of Francis Bacon … "*Man prefers to believe what he prefers to be true.*"

To underline the main message here, think of a very young child in a sweet shop without any supervision. We can easily imagine their childish motivation to sample a few of the items on display that are within their reach. They would probably do this without any sense of guilt or real understanding that what they are doing is wrong. It would appear to be a similar situation in the worlds of theoretical astro-physics and theoretical mathematics. It seems to come down to simple human nature that, in the absence of a perceived need to follow some form of professional code, and, while effective external controls are non-existent, some people will tend towards distinctly self-serving behaviour, often unconsciously. In other words, what we might describe as people with no constraints 'letting their imaginations run riot' while not necessarily being focused on reality or being fully aware of their actions and the consequences thereof.

Take this idea into the situation where we have working astro-scientists today with their intellectual freedom and their freedom to act, as they see it, being enshrined as a right. It does not matter where we find these people positioned on the professional career ladder or what they actually do within a particular hierarchy, the same will apply. Accepting first that many scientists, if not most, will feel a strong, fundamental need to be true to the tenets of their chosen profession, there are undoubtedly others who are not focused in this way. These others may, in order to achieve and maintain a quiet life and a quasi-successful career, take the view that they have no choice but to stick with broadly accepted theories and to just get on with non-contentious research and ultimately minimally productive work. There might also be those who put materialistic tendencies behind them but who wish to direct their energies single-mindedly towards protecting their positions in a hierarchy of their peers in order to stoke their own egos. Others might find themselves with an involvement in ongoing prestigious work for which they will do anything to protect said project's status and ongoing funding. Some will put their families and economic situations first and will just not speak up when involved in work that they view personally as questionable. Some others, sadly, will have a clear awareness of what the right things are to do but lack the courage or moral fibre to break ranks and openly state their doubts. Through whatever form of inaction we care to consider, any perpetuation of pointless work and waste of public money is just not acceptable [2-1]. Shame on those who fit any of these definitions, for every one of us is 100% responsible for the decisions we make. People are not born to be scientists, they choose that vocation because of their environment and education as a profession that comes with responsibilities to science itself and to everyone. By not challenging things where it is right to do so, scientists are serving an already damaged system and they are intelligent enough to know better. No matter the reasons here to ponder, the major variable that clearly remains is that of basic moral judgement. It is an uncomfortable thought to consider that it is quite possible for conformists who are both weak of character and yet arrogant within these definitions to be the ones who, by saying and doing the right thing at the right time, can rise to positions of influence within the astro-science discipline. Is there not then a danger that this type of person will go on to dictate the way ahead for others and for astro-science research overall? It is a fact that, until a major shake-up comes along, where people find out they are only numbers within a weak system, conformity will be rewarded by those systems. Sadly, a mixture of these unprofessional attitudes has already created a mess for the up-coming generation of young and enthusiastic scientists to deal with and to sort out. This is my analysis of what is likely going on within astro-science; a situation that is perpetuating the confusion and lack of clarity we are currently experiencing.

In total contrast to this, we must remember there are scientists for whom a focus on the right things for the right reasons is the natural way they operate. These are people whose lives and careers are in balance. I am confident in suggesting that the majority of scientists are or want to be like this. We also need to remember, however, that the up-and-coming generation of young scientists may find themselves under pressure from non-performers above and around them to conform to situations that stifle their own independent thinking. For these good young folk to get ahead, especially in the current intimidating world of established astro-science research, they could find themselves strongly pressured to comply with mainstream views, with no obvious option to do otherwise. What a waste! The big questions that arise from this are, "*How is this array of unfortunate attitudes affecting people right now?*" and "*Why has science gone along with this situation for so long?*"

To begin exploring some of this and to bring into better focus the story of our universe, let us again go back a hundred years or so, to the time when there seems to have been a split in the world of scientific research, one that was fundamental to the explanation of why we now have unproven and currently un-provable theories being supported as scientific fact. This split came about when research developed to the point where there were some scientists choosing to work on gravitational theories and others choosing to work on electrical/plasma theories. Things had previously been good because there had been open interdisciplinary cooperation between all areas of research - in fact, science had previously been known as 'Natural Philosophy'. These two quite different fields went their separate ways during the early decades of the 1900s. Gravitational research, based as it was around the respected work of Newton and Einstein, forged ahead with the support of the reputations of Einstein's eminent followers who had developed an understanding of his theories for their own purposes. As a side-effect of the limelight directed at Einstein's work and that of other gravity supporters, the discipline of electrical/plasma research found itself in the shadows. It fell behind with its body of scientific support and the involvement of motivated graduate students to become an area that seemed mainly to serve the needs of the commercial world of power generation, distribution and associated manufacturing applications. The result was that the mystery-riddled world of Einstein's gravitational theories became adopted 'as was' by astro-science and thereafter supported in significant and high-profile ways. Success was thereafter guaranteed for those who got involved, for it all seemed to make sense. This meant that the less understood discipline of electrical/plasma science had effectively been painted into an industrial/commercial corner, there to remain for many decades.

We will take another step back to the early 1900s to pour a little more concrete around the base of this story. This was when the respected astronomer Sir Arthur Eddington carried out a telescope observation experiment from an island off the west coast of Africa. The 'successful result' of this experiment seemed to prove Einstein's theory which said that light would be bent by the influence of gravity. A successful result was interpreted when Eddington observed light from a star that was actually situated behind the Sun to his direct line of sight. The Sun's gravity was therefore judged to have bent the star's light around it.

Eddington's experiment © author

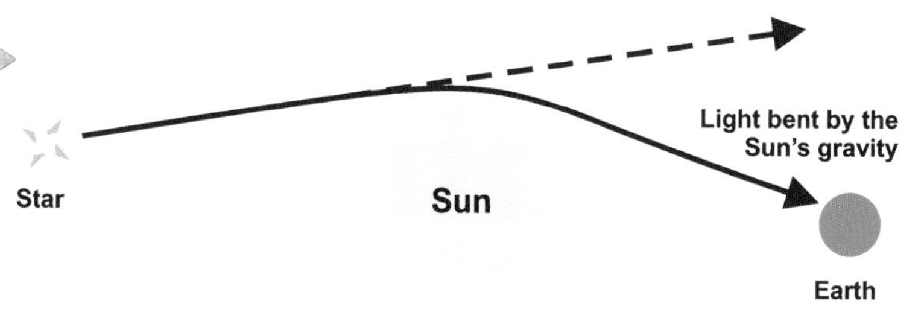


This finding delivered a great amount of credibility for Einstein and his work. It seemed that at last, something practical from real-life observation supported the major claim about gravity in Einstein's general theory of relativity.

In addition to this, and just as important to our story here, Eddington was about to come up with a theory that would be taken as answering the important question of how stars work; our own Sun included. The story previously discussed was that stars were burning balls of gas which, due to an internal pressure being created through some unknown mechanism, were able to produce an outwardly-acting force equal in all directions. This force was viewed as a balancing force, adequate enough to counteract the inward pull of the star's own gravity. This was the partially complete model adopted as the explanation for how all stars were able to maintain their ball-like structure. However, the internal mechanism that allowed this process to occur was the big problem to be explained, but as far as it went, this description was good enough to kick things off.

Eddington's refinement was to add a 'nuclear explanation' to explain how energy was being generated at the Sun's core. Conveniently, his contribution to star operation theory came out around the same time that discoveries were being made in the realms of atomic structure and nuclear fusion and fission. Fission reaction (atoms being broken apart and lots of energy being released as a result) is what takes place in the cores of our current nuclear reactors. Fusion reaction (atoms being forced together and lots of energy being released as a result) is what Eddington claimed was going on in the core of our Sun, and by association, all other energy-producing stars.

The process of nuclear fusion is now broadly accepted as taking place at the core of stars as the source of power generation that allows them to radiate energy. This occurs through gravity's compression of hydrogen gas producing a self-sustaining element conversion process that generates great amounts of energy in opposition to the inward pull of a star's gravity. This conversion process involves hydrogen atoms being combined to form atoms of the heavier gas, helium. During this process, sub-atomic particles that are surplus to requirements are released as heat, light and X-rays. This fusion process is now the widely accepted explanation for the internal operation of all energy-producing stars. Keep in mind, however, that the fusion process comes with the implication that the hydrogen gas all stars are said to be made of will eventually be completely used up. When this happens, the star will die. This of course would also apply to our own Sun, but fear not, it is common knowledge that there are very significant problems with this theory.

Getting back to Eddington's contribution to the theory of how our Sun works, the missing energy process employed in the central furnace had apparently been identified as nuclear fusion. And since this was a process that could, theoretically, be initiated by the force of gravity, everyone involved was happy. This more complete theory has over many decades been accepted as the truth of how our Sun and all stars function. We will see about this later.

I have focused on stars, purely as a vehicle by which to highlight the issue of how easily ideas and theories can be brought into being and automatically accepted by people. Other questionable theories with a gravity association are equally supported as scientific fact by the astro-science establishment. It is important to realise that the acceptance of all these things has been influenced greatly by a respectful view of Einstein's work and the assumed existence of a credible astro-science establishment that is headed in the right direction. What kind of brave person would ever be prepared to argue with something as formidable as that? Further to this, the apparent acceptability of the SM and its gravity theories has had a further quiet boost, this time through nothing more than the passage of time. The whole situation has become normal to us through nothing more than it just having been around for a very long time. What basis could this ever be for establishing scientific fact and for making the real progress we need?

# 3 | We are waiting for answers to these questions

In chapter one I mentioned some of the issues we have with the standard model. I will expand on these here in summary form to suggest that many of them have arisen from adventurous assumption and a myopic approach to a particular line of theory throughout the 1900s. I will then add a few issues that come from more recent findings. The intention is to show clearly the reason for the state of great puzzlement that a growing number of scientists, researchers and interested members of the public feel these days about the attitude displayed by astro-science when issues that confront their pet theories are raised.

Astro-science has been almost halted intellectually by fundamental questions for which it cannot produce good answers. The typical response is to ignore, or with the wielded hammer of their superior intellectual position, belittle or even attack those who present these questions. Meanwhile, the world moves on and other scientists who do not suffer from constipated thinking are building a case around the questions in play. As you progress through this chapter, please do not feel you must understand all the detail; explanations for many of the basics will come later. The objectives of this chapter are threefold: to act as a consolidation of the main issues; to highlight the serious implications they represent; and to open the door to alternative thinking.

---

Dark matter and dark energy are said to account for the structure of the universe and validate the claim that it is expanding. There is much room to wonder about this …

First, in the case of spiral galaxies, we are assured that the total gravity produced by conveniently placed dark matter, plus the galaxy's normal matter and that of the supposed super-massive Black Holes at their centres, is enough to explain how they hold themselves together and behave as they do. Why then does a typical spiral galaxy appear like a revolving solid disk of stars rather than the stars in its outer regions being seen to orbit more slowly? If gravity really is in play here, then given the way its influence is shown to reduce with distance, spiral galaxies should revolve more like froth does on the surface of your coffee as you stir it in your cup. It would be fast on the inside and slow on the outside, with the outermost stars even trying to fly away!

A larger scale contradiction also exists here. On one side we have the claim that everything is flying away from everything else and that the universe is expanding, and on the other side we are told that some galaxies are actually colliding! When this was first found, astro-science pleaded ignorance and did not address the issue, for they could not, and despite all the plausible mumbo-jumbo that you hear, they still cannot. In addition to this, why do we observe groupings of galaxies that appear to be connected by thread-like structures? Surely, no groups of galaxies would exist if all matter was initially dispersed by an enormous Big Bang explosion!

We looked earlier at dark energy as an 'anti-gravity' invention that helped account for the apparent expansion of the universe. Well, it has additionally been associated with the claim that this supposed expansion is accelerating. The explanation for this starts with the rapid reduction in the gravitational influence produced by dark matter as galaxies and all matter moves apart. This rapidly diminishing 'backward-pull-force' is claimed to be constantly opposed by a 'forward-push-force' provided by dark energy, which unlike gravity, we are told must diminish at a slower rate over a given distance. The result is that an overall and ever more effective 'push' is supplied to all things by the dark energy force. This is a simple version of the explanation for the acceleration that astro-scientists believe they are seeing. Here we have a case where one piece of imagination is being put forward as proof that another piece of imagination is real and actually works! This is an example of the same type of unacceptable 'idea-interdependency' process through which much of the SM 'house of cards' has been built.

Things just kept deteriorating for the gravity model, and another sticking plaster was required when better observation data revealed a further problem. Astro-scientists had judged the dual narrow beams that shoot out for millions of light years in opposite directions from the cores of galaxies to be jets of high-energy particles. Wait a minute though, how could anything even begin to leave the core of a galaxy, if as we are told, the gravity generated by the super-massive Black Hole at its centre would not allow anything to escape from there, not even light? Moreover, how could these tight jets of gaseous matter hold together so well in their long and narrow form in the vacuum of space? No gas would behave like this in a vacuum, it would disperse very rapidly indeed!

To answer these points, astro-scientists suggested, very interestingly, that the involvement of magnetic fields would explain the ionised gas being held together. They had come up with what to them was an exotic theoretical reason while avoiding acknowledgment of the simple fact that electric currents are required to generate magnetic fields! The simple fact, well known to electrical engineers, and one would think to astro-scientists as well, is that electric current flow is first required for magnetic fields to exist. Not however being trained in electrical or plasma science, it seems our current astro-science community does not appreciate this fact.

The idea of a galaxy emitting 'jets' from its core © author

Today we are told that the make-up of the universe is 4% ordinary visible matter plus 96% invisible matter. The invention of these dark, undetectable components has been challenged since they were suggested. So as far as their relevance to this book is concerned, they contribute nothing, so we will say no more about them.

As I hope to show, new evidence indicates that the structure and operation of our universe relies on something very different from the explanations provided through standard gravity theory. It seems instead that it works on the logical cause and effect associated with a force that established science already understands very well; electricity, or to be more correct, the electromagnetic (EM) force.

Centaurus A displaying its 'jets'
Credit: X-Ray NASA CXC CfA R Kraft et al

The next thing to mention is **Inflation**. This is the name for the process said to have taken place just after the Big Bang. Among other things, it is offered as an explanation for the smooth background 'wallpaper' of space known today as the Cosmic Microwave Background (CMB); this often being described as the furthest away 'thing' in the universe that can be detected. Inflation was mainly invented to explain the smoothness and close temperature relationship of everything in every direction in which we look. It offers nothing to help us with explaining Electric Universe theories so I will only touch on one of the more obvious issues it was invented to solve, the 'Horizon Problem', plus one other that seems to involve 'not enough time'.

The Horizon Problem is a situation that seems very odd from here on Earth. It becomes apparent when we do two things: (1) Look at two opposite furthest observable points in any direction in space and take temperature measurements at those points. (2) Remember that the distance to any furthest point is what gives us the figure of 13.7 billion years as the age of the universe. Here, we find that the two measured temperatures are within a whisker of the same value, but that the calculated distance between those points means that they could never have been in the same location in the past where they could have attained the measured similar temperatures. In other words, if it is correct that our universe did start at an infinitesimally small point known as a 'singularity' and that the maximum speed of light is 300,000 kilometres per second, then any such distant opposite points are just too far apart for the available 13.7 billion years to account for.

A further significant issue about estimated distance and universe age is that a galaxy cluster named COSMOS-AzTEC3 [3-1] has been observed through standard reckoning to be 12.6 billion light years distant from us. How could a fully formed structure like this have had time to assume a mature state within the 1.1 billion year time difference between its measured age and the 13.7 billion year age of the universe?

This is just not possible to reconcile logically, even on the basis of astro-science's very flexible theories. The bottom line is that they continue to rely on the idea of inflation and its process of rapid expansion at super-luminal (faster than light) speed, so their message to everyone can remain to be that everything just got really big, really fast. It apparently does not matter that this brave claim contravenes their hero Einstein's universal speed limit; a theory, which it must be said, is being seriously questioned today. Astro-scientists seem prepared to come up with just about anything to protect what they have intellectually, and in many other ways, invested in. To a great extent they get away with doing this, so in the public domain at least, we still have the idea of inflation as an accepted fact. Are we really seeing as far as we have calculated and assumed we are able to? Is the universe actually expanding for the reasons we have been told? Can we dare assign an age to the universe based on current theory? I think that if we stick to the basic evidence and not let ourselves be overwhelmed by complicated and confusing reasoning, we will see that we still do not know enough to say too much at all.

In opposition to the standard story about **Quasars**, it seems they are not the oldest and furthest away high-energy emitting objects in the universe. There is a wealth of observational data to support claims that many of them are actually relatively young objects joined by 'bridges of matter' to active galaxies, some of which are quite close to our own Milky Way galaxy.

◀ Quasars: young, small and hot to older, larger and cooler © author

Quasars are, for their apparent physical size, found to emit exceptionally high levels of radiation and to give off intense light. Astro-science has traditionally used an interpretation of the redshift of this light in ways that prove to their own satisfaction the distance these objects are away from us. However, this interpretation has been shown to be highly dubious. Quasar emissions are actually measured at distinctly separate levels of energy (redshift) and not as values on a smooth linear scale. To astro-scientists, a linear scale would be preferable because they believe that quasars are moving steadily away from us and that they are close to the edge of the visible universe. Questions are therefore raised as to how these distinct steps in emission energy are being generated and what they represent. Furthermore, since quasars are often found close to the cores of active galaxies, might there be a reason related to the powerful forces in those cores that could better explain the existence of those quasars? And why is it that quasars can be observed along lines leading away from active galaxy cores?

**Stars of various types** … Protostars, T-Tauri Stars, Main Sequence Stars, Red Supergiants, Blue Supergiants, White Dwarfs, Yellow Dwarfs, Red Dwarfs, Blue Dwarfs, Brown Dwarfs, Black Dwarfs, Cepheids, Neutron Stars, X-Ray Stars, Pulsars, Magnetars, Quark Stars, Preon Stars and Wolf-Rayet Stars are potentially confusing when considered as different types of bodies. In fact, we should take this array of names as an indication of the embarrassment of ideas about stars that really does exist! They are often talked about in the public domain as if their various labels should define them as very different objects with individual explanations for their creation and operation. But we can relax here, we do not need to follow this line of thinking if we just want to understand a plausible explanation for why all visible star types appear as they do.

Here is a brief reminder of what we recently learned about the currently accepted story for stars. They are said to result from the process where gravity brings great amounts of gas and other heavier matter together. When everything about this task is complete, with a great amount of energy assumed to be released in the process, a star will have been formed of a particular size and energy output. After this, we are told that a clock starts running that defines a lifetime for the star; a period during which it passes through a sequence of stages represented by some of the names mentioned above. At the end of an active star's life it dies and collapses in on itself when the gas (fuel) it is composed of runs out. This process of collapse is termed a nova or supernova event; the distinction here depending on the original size and mass of the star involved. These nova events are where some assume it is possible for the matter thrown off by an exploding star to be accelerated to great enough speeds that its subsequent mechanical collision with other particles of dust and gas in the vicinity, will produce high-energy radiation such as x-rays and gamma rays. Because of problems that exist with theories on which this account is founded, it is the story we will be challenging. In chapter six we will look at star creation from an alternative vantage point that embraces the theories of powerful electrical forces we already understand.

I do not want to focus too much on the questionable aspects of the standard model and I do not want to be seen as documenting only an attack on these things, for as you will see, the book has a much more constructive objective. It remains important, however, to show that a very complicated and confusing story has evolved to include narrowly applied theories and a lot of assumption, the origin and operation of stars being just one aspect of this. Amusingly, and in terms of how the gravity model has come together, I am here reminded of the old Johnny Cash song … "One Piece at a Time" … The lyrics of this song tell of a car being built from 'opportunistically purloined' parts over a period of time. The end product is a rather special contraption; a vehicle which I choose to think would only have been of great value to its builder. In a sense, this is an analogue for how the SM has been put together. When it is discovered that something else is required to patch it up, then that 'item' is imaginatively constructed and fitted into place.

**Black Holes** are 'real objects' only in the imagination of supporters of the gravity model. The supposed process of creation of these things is similar to the one involved with the death of typical stars, except it is on a very much grander scale. The need that brought them about was to explain observations that could only be attributed to the effects of assumed, colossal gravitational forces. The case made by astro-science for the existence of these 'gravity monsters' has been built on a foundation of unrestrained mathematics and is another good example of a theory that everyone has just accepted. Black Holes are components of the gravity model which, in themselves, lend nothing to the purpose of this book. Their explanation has had 'extra bits' added as required, so the overall story around them has become complicated and confused. Not many people understand what is actually being described and any information that does appear tends to be of a sensational science fiction nature. I will say more later about the forces and processes that have led astro-scientists to believe that Black Holes are real.

I will leave commenting on theories for a moment to say something about the information given to us through the collaboration of astro-science with various parts of the media. Although interesting and highly convincing, many of the documentaries, magazines and books that have been produced to explain ideas from astro-science only present a narrow, superficial view of things. It is true that if we look further we are often put off by the complexity we find, so we tend to quickly stop looking and thinking about these things. However, it seems that our universe is less complicated than currently portrayed, so in what lies ahead, I want to introduce the alternative and simpler theories which for many of us bear this out. I am led here to think of the wonderfully simple visual education experience that could be produced around these alternative theories, if even just a portion of the funds allocated to astro-science documentaries and professional presentations was made available.

We all too frequently find ourselves taking a difficult route when trying to understand things or solve problems, only later to find that we have disregarded the simpler and more obvious routes that should first have been considered. This was a concept known and understood well by a particular English Friar and philosopher of the 1300s, William of Ockham. He supported the notion that when looking for solutions, we should not make unnecessary assumptions and should initially consider the more obvious and simple options. Today this idea is summarised under the label 'Ockham's Razor', and in terms of the current potpourri of complicated theories that astro-science has spent its time on, it certainly seems applicable as a caution to heed.

The **thermonuclear model of the Sun** also has many problems, mainly because the Sun does not behave in the way astro-scientists assume and tell us it should. The Sun has provided obvious clues about the way it works for many decades but the questions that have arisen from these clues have, unsurprisingly, not been satisfactorily answered. The situation is embarrassing, and it is a rather odd one for astro-science to continue to defend, as they do. Why is this embarrassing? Well, it is because we are being asked to believe that the Sun operates in a way that can be described as both a severe stretch of the imagination and a scientific impossibility! Here, astro-science's attitude to the thermonuclear model of the Sun cannot be called an open-minded one. Despite great amounts of evidence that indicate better alternative theories, astro-scientists seem strangely unshakable in hanging on to their outdated ideas and unworkable theories. A far more plausible explanation for the Sun's operation is available if we include and consider things from electrical and plasma science with which we are already familiar. From that new vantage point, we are able to take a fresh look at how all stars operate.

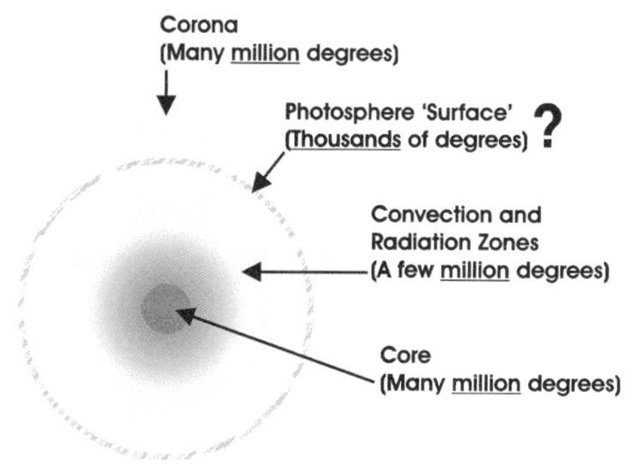


Here are just some of the issues we have with the current thermonuclear model of the Sun:

Why is it that the Sun's 'nuclear core' temperature of around 15,000,000˚C drops to around 6,000˚C at the surface then increases again to 2,000,000˚C (and beyond) in the Sun's surrounding corona?

The temperature profile we are currently given of our Sun © author

Solar astronomers have been working for decades to answer this question and the absence of anything convincing speaks volumes about the basis on which they have looked at it. They have not even come close to convincing serious-minded folk with what they have come up with, and they remain puzzled. Amusingly, I have seen this temperature profile problem described as 'getting warmer the further you walk away from a fire!'

Then we have the issue of **Sunspots**. These are described as being caused by magnetic fields leaving and entering the surface of the Sun. The fact that the centres of sunspots are found to be 1,500˚C cooler than the surrounding photosphere is typically ignored in explanations that include talk of 'tangled magnetic fields'.

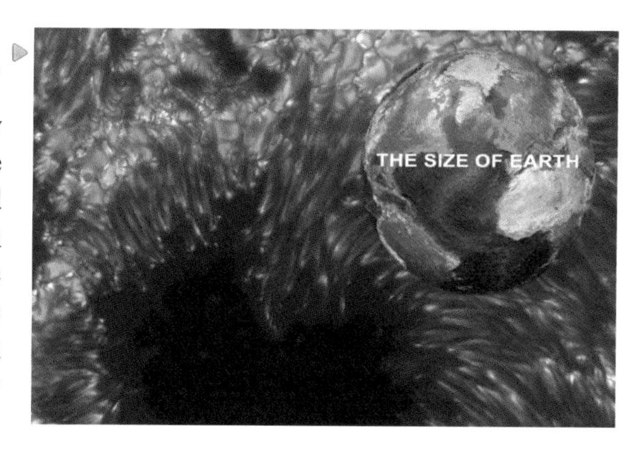


A Sunspot compared to the size of the Earth (adapted image)
Original Image Credit: SST Royal Swedish Academy of Sciences

They say these tangled fields stop an amount of heat energy from inside the Sun getting to its surface. This is extreme conjecture and a testable theory that has never been proved workable. Astro-scientists get in a muddle and tie their own hands by avoiding discussion of the simple fact that electric currents are required in order to generate the magnetic fields they so often rely on in their explanations. How can we make progress when they are so selective in what they choose to consider from existing knowledge?

Next we have the **Sunspot Cycle**. This is where our Sun has been observed to go through similar patterns of sunspot production in approximate 11-year cycles. Astro-scientists have repeatedly tried to explain this through complicated magnetic fields affecting one another. Here they have limited themselves by thinking only of things from inside the Sun; they have no framework whatsoever for considering forces that exist outside the Sun.

They have therefore failed on every occasion to produce a convincing case but have nevertheless been enthusiastically encouraged and funded to keep going with their investigations. The issue here seems to have its roots in the fact that astro-scientists see everything that is energy production-related as coming from inside the Sun and take no heed of the electrical forces in the Sun's environment that actually do influence what they are observing. What we end up with is the same situation where we put a set of blinkers (blinders) on a horse! This being the case, I hope to show that the notion of external electrical effects is the key to understanding and answering many of the questions posed by the Sun's apparent behaviour. By achieving this, we will also be well placed to consider the operation of all stars.

**Redshift**, in relation to light coming towards us from objects in space, has traditionally been used to tell us how far these objects are from us and by how much they are accelerating away. I mentioned previously that the originator of this particular idea, Edwin Hubble, ended up having his own doubts about it. His further analysis that led to his revised position has been ignored, and so his good name is still firmly but inappropriately associated with the original use of the term.

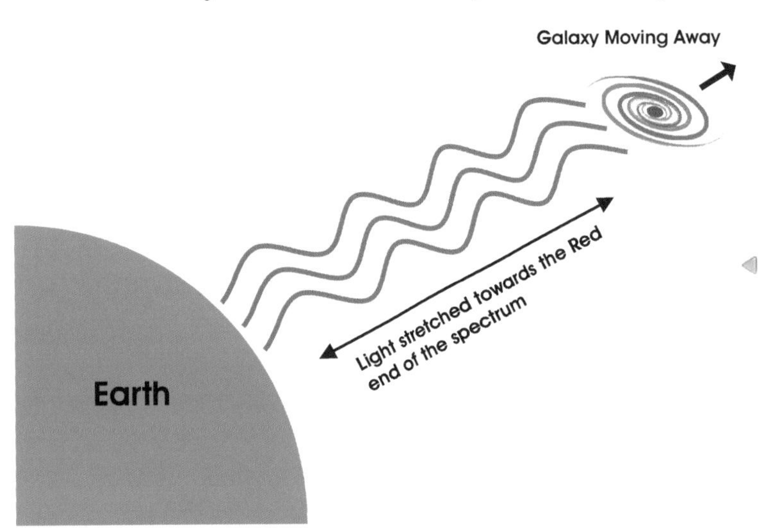


How light emitted by an object moving away is stretched out © author

One of Hubble's students, Halton Arp, also came to realise that there was more to the subject of redshift. As part of his work he developed a database of information that seemed to prove that the assumptions held about redshift in those days were wrong, and by implication, the many calculations that had been based on it were wrong as well. This of course would have been devastating news for the astro-science community if it were to be widely talked about, so 'the establishment' had to make Dr Arp's voice on these things irrelevant. This came about in a very effective way, as you will read about when I say more about Dr Halton Arp in chapter five. Nevertheless his evidence, which still stands today as proof that the standard interpretation of redshift is wrong, has been re-confirmed since and waits in the wings to be acknowledged when the mainstream's more enlightened astro-scientists see it for the valuable contribution it is.

The focus of Arp's interpretation of redshift is not on the distance or speed of motion of objects being observed. He found instead that redshift actually has two component parts to it, each with their own redshift values.

One value does indeed relate to the motion of the object, and the other **much more significant** value, relates to the object's age. It so happens that both these values have traditionally been taken together as the single redshift value from which all the motion characteristics of deep space objects have been determined. This has obviously been an invalid step to take because the redshift component that relates to the object's age, termed the 'intrinsic' component, should be subtracted from the total before any motion calculations can be performed. Nevertheless, astro-scientists have continued to take no account of this, so an astounding situation exists. This is why the results of their calculations cannot now be taken as proper evidence to support anything they have claimed to be true to this point.

The further important but sad part of this is that the conclusions drawn from these calculations are still put forward to support well-entrenched claims: the age of the universe at 13.7 billion years; its ongoing expansion and the Big Bang's moment of creation. These grand inferences now all appear to be wrong. I hope later to show that the intrinsic redshift component, most often mentioned in association with quasars, is actually an inherent property indicative of their recent formation. I will go on to suggest that the fantastic amount of redshifted light energy they are judged to radiate is actually an indication of their age and that this has nothing to do with distance or speed of motion.

It seems that a proper analysis of redshift would threaten the house of cards that has been built around it. This is important to mention because it is yet another example that highlights the attitudes that have evolved to be widespread throughout the astro-science establishment. What seems therefore to be the case is a situation where exists an open door to questionable theories; the resulting environment also likely influencing the decision-making hierarchy that determines the future direction of research and education. I would venture to suggest that many good scientists see this going on and regret it deeply. I remind you here of the quote I included in my introduction to this book; a statement made in 2006 by Wallace Thornhill, physicist and major advocate of the Electric Universe model. The quote comes from his website article "A Real Theory of Everything" [3-2]

*"We have to discard 'modern' physics and return to classical physics of a century ago. This, perhaps, is the greatest hurdle – to discard our training and prejudices and to approach the problem with a beginner's mind."*

This is a powerful statement that has implications beyond astro-science. It takes courage to say something like this, for it seems that if you do, you are kissing goodbye to any hope for a successful and comfortable career in the mainstream. Nevertheless, various honourable people have done just this; they have acted on firmly held convictions and I think, therefore, deserve our respect.

**Comets** were first described in 1950 by the American astronomer Fred Whipple as dirty snowballs melting in the radiated heat of the Sun. They have now been closely inspected by sensitive measuring equipment carried on space probes, and some being pointed at them from the ground. To those with open minds who remain receptive, the data returned tells us that comets are not the dirty snowballs that astro-science would have us believe. And in terms of how many there are, despite the problem of coming up with a figure in the first place, an estimate of around 6 billion was originally statistically predicted.

◀ The two tails of a typical comet nearing the Sun © author

This number was found to be far too low when better data analysis was conducted and computer simulations were carried out, so it was raised to a figure of ~400 billion. Overall, this has turned out to be a big issue for cometary scientists, because there now seem to be far too many comets to suit the preferred story about their origin and the calculations used to come up with predictions about them. To address this and accommodate the larger figure, ideas were revised to claim that beyond those left over from our own solar system's formation, the Sun must have captured most of its comets from other stars billions of years ago when those stars came close to our solar system. This is quite a wild piece of speculation and it has no basis in anything whatsoever. However, being just another guess from their limited gravity toolbox, it highlights yet again that sticking plasters are quite happily applied by astro-science to patch up their theories. The sad thing is that it is mostly the public purse that funds the highly questionable work of mainstream astro-science and that currently at least, most of those involved enjoy well-paid jobs, rosy futures and the prospect of nice pensions.

Digressing for a moment, there are a couple of things to underline. University students entering astro-science disciplines should not be faced with this type of attitude for it may influence the direction of their careers. A few may be tempted to go along, but I prefer to think that most others would take a more responsible and independent approach. It is also apparent that the public shows little interest in science in general, for many of us do not see much within it playing a significant role in our daily lives. However, reasons do exist for us to consider this attitude carefully and for us to think more about what is going on in science research; outcomes from which have already determined much of our own and our children's futures. If we were to take a moment to lift our heads out of the daily grind, we would be giving ourselves a chance to realise this is very important stuff and that we should be paying more attention to it.

A directly associated issue is that comets are seen to come into the solar system from just about every direction in space. Because of this, astro-scientists now assume the existence of a 'shell-like globe' of lumps of ice that surrounds the whole solar system and that from within its population the origin of comets is likely very high.

These assumptions have backfired somewhat and have generated further questions. The claims that comets are left over fragments from the formation of the solar system and that they can provide us with important clues about its origins have for some time been contradicted. This is because we are now told that comets could have been captured from other stars, a claim further added to by the statement that these comets now provide an opportunity for us to understand the likely composition of other stars. This makes me wonder how much wild speculation is being practiced and how much today's astro-science is being allowed to get away with. I hope you get a feeling for the easy willingness of astro-scientists to play with and alter their theories when unexpected new evidence and unwelcome problems come along. This is becoming a predictable model of behaviour today, especially among those who revel in the opportunity for personal exposure, either on TV or through other forms of media or through the books they write. This should, at the very least, make us wonder about the quality of information that has been supplied to us as 'trustworthy and believable'.

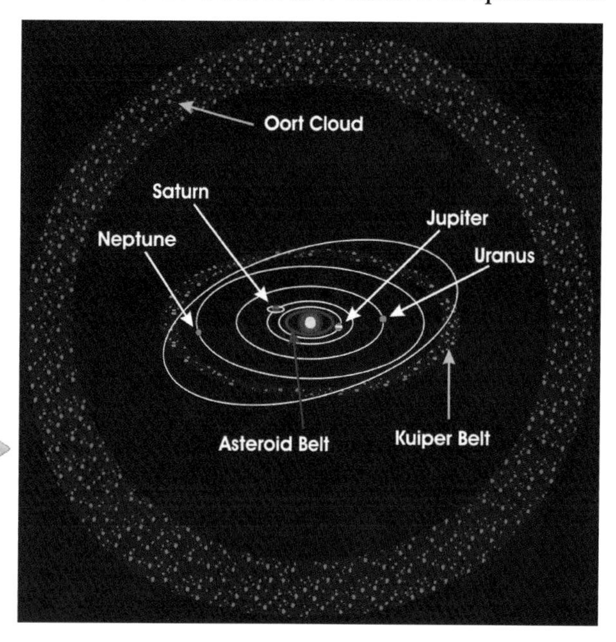


The Oort Cloud, Kuiper Belt and Asteriod Belt in relation to the Solar System © author

Another claim of the standard comet model says that they are influenced to leave their stable orbits in the far away location of the 'Kuiper Belt' that is situated beyond the orbit of Neptune, or in the even further away Oort Cloud, by the gravitational influence of passing stars or through collisions with other companion bodies. However, some astute astronomers have observed that if this were true, then we should see many more comets in groups and not as the single visitors we do. To help picture here the unlikely nature of prediction versus actual observation, I will raise again the subject of scale, for it is important that we have an idea of the kind of distances and probabilities we are dealing with.

I will give an easy to picture description regarding this model, one that was originally put together by astronomer Tom Van Flandern and which I saw quoted by Wallace Thornhill in one of his many excellent articles about comets, "Deep Impact 2" [3-3] on his website.

*"If the Earth's orbit were represented by the period at the end of this sentence and Pluto's orbit by a circle one centimetre in diameter, then the nearest star is 41 meters away. The Oort cloud of comets would orbit in a sphere 6 meters in diameter containing one comet per cubic millimetre. The comets would move about 3 millimetres per 1,000 years. They are effectively motionless. Passing stars 'whiz' past at a meter per 1,000 years and stir up the nearby comets. Less than 1 in 10,000 disturbed comets will be knocked onto a path that will target the 1 millimetre or so sphere surrounding the Sun where a comet might be seen from Earth."*

This is not suggesting we should not see comets here on Earth, because we do. Rather, it is a comment on the fact that we observe so many, but not in showers.

When listening to experts, they often refer to the tails of comets being, in part, brought about by heat from the Sun that forces 'melting' (sublimating) ice, gas and dust to rapidly leave a comet's surface and then for it to be carried along in the 'solar wind' as a long narrow tail behind it. This turns out to be a narrow view because it is based on old assumptions that have never included any consideration of the electrical nature of space. As far back as the 1800s, scientists and science magazines are known to have announced the distinct possibility that comet displays were fundamentally electrical in nature. It seems to have been for popular gravity-centric reasons that this line of suggestion was ignored in favour of the dirty snowball idea. It also seems that things are destined to go full circle with this issue, because electrical activity on and around comets has now been proved as fact but still remains publicly unrecognised as such. If the Sun's heat and the solar wind are responsible for comet tails, then why in 1996 was the tail of comet Hyakutake detected by the Ulysses spacecraft mission to be 360 million miles behind it? ... Four times the distance that the Earth is from the Sun!

You will be aware of how hair can be made to stand on end when attracted to a nylon comb by the electrostatic charge built up when you stroke it through your hair. With this impression in mind, consider the thin bright jets seen here shooting out from the surface of comet Hartley.

Comet Hartley displaying its "jets"
Credit: EPOXI_NASA JPL-Caltech UMD

These jets are explained conventionally as coming from narrow holes in the surface of a comet that act as 'gun barrels' to direct jets of sublimating ice (as gas) from under a comet's surface out into space. It would be strange if this were true, because similar jets are observed on the cold sides of comets that face away from the Sun. It seems instead, they are rather like the appearance of hair when it is electrostatically attracted to a comb, so perhaps there is a more likely electrical explanation for what we see.

Another thing about comets is the vast 'glowing cloud' often observed surrounding them as they get close to the inner solar system. Again, the dirty snowball idea tries to explain this cloud as sublimating ice firing off gas and dust into space, where the solar wind then interacts with this material to make it glow in a form that we refer to as a comet's 'coma'. However, when more than a fraction of a second is applied to thinking about this idea, it appears to be quite silly. This is because we not only have firm evidence that proves the ridiculous nature of the suggestion, we also have common sense and logic on our side.

This is highlighted best when we consider that the coma of comet 17/P Holmes in October 2007 brightened by nearly a million times in just a few hours and that it was measured to be 2 million kilometres in diameter ... this is a cloud that is larger than the Sun! The crazy thing was that the nucleus, being the actual solid core of comet Holmes, was itself only 3.4 kilometres in diameter! An obvious question therefore comes to mind ... what could cause a practically instant stellar scale effect to emanate from a tiny rock? ... Might the answer be electricity?

Comet 17/P Holmes - Creative Commons Image © Gil Estel

Then we have the observation involving comet Halley's 'flare up' in 1991 when it was travelling between the orbits of Saturn and Uranus. This being a distance some 14 times further away from the Sun than the Earth is, there is no appreciable heat at that range to account for such a violent display, so again, what could have caused this?

We are able now to inspect the surfaces of comets in good detail by way of much better resolution imaging and detailed sensor data being returned to us from spacecraft missions. These data indicate clearly that the activity on a comet's surface is not brought about by sublimating ice but instead by a very different process. Their surfaces are often seen to burst into a high-glow mode, an event that often signals the beginning of a rapidly expanding coma. This visual clue usually appears quickly and is typically explained as being caused by large surface eruptions of sublimating ice. But as we have said about comet Halley, this behaviour has been observed far out in the solar system in a very cold region of space indeed.

The rocky and sharply defined surfaces of comets, along with the craters on their surfaces, and now the mini-mountains we can see, are all whopping great puzzles to explain if comets really are big lumps of ice and dust. Another view of this can be formed when we consider something we already do routinely here on Earth. This is where we use powerful electric currents for a number of purposes in our manufacturing processes. Here we have things such as welding, cutting and material removal from solid components.

What if this powerful process based on the electrical removal of material from the surface of a body was scaled up by nature and applied to the surface of a comet? Could that process, in a vast electric environment, possibly be the explanation for the creation of the sharply defined features we see on comet surfaces? Keep in mind that what we know has already gone on in space over eons is restricted to a period of only a measly hundred years. The temperature of some comets has also been measured to be hot, not cold, as you may have expected. Areas of the surface on some comets have also been found to be very black indeed and to display behaviour just like that of the exceptionally fine black soot produced by a short-circuit that causes electrical burning in a domestic situation. Many comets have also been found to give off extremely high-energy radiation that could never in a month of Sundays be accounted for by the dirty snowball idea. Moreover, heavy elements (minerals) have been found in the gas and dust expelled from comets, where the production of these is impossible to achieve with the temperatures and compositions involved in the icy comet model. Now that we can get up close and personal with comets by way of the probes sent out to inspect them, we are justified in believing that they hold a tantalising and imminent promise of disproving, once and for all, the tired old dirty snowball model.

We can see **Craters** on the surface of every body in the solar system, except, of course, for the Sun and the gas planets. To account for the rocks of various sizes and compositions that are supposed to be responsible via impact events for creating these craters, we are provided with a variety of explanations as to their origin. The most popular of these is where planets and moons have collided and broken up in the dim and distant past and where big rocks like asteroids have smashed into the surfaces of planets and moons with enough force to eject great amounts of smaller debris into space. Whatever origin is claimed for this menagerie of various sized solid projectiles, in the minds of today's astro-scientists, these objects prowl our solar system till they find, through gravitational connection, bodies to collide with and leave on their surface the 'signature impressions' referred to as impact craters. The energy supposedly delivered in producing those craters is known in engineering as 'kinetic energy'. This is the class of energy released when a body travelling at speed hits another body. On the surface of it, this explanation seems to fit, but as I said before, have we really thought much about these?

I must mention also that in terms of the gravity model, the formation of planets and moons is given to us as being based on the idea that countless numbers of floating rocks gradually collected and fused together under the influence of gravity. If we ignore the certain lack of required kinetic energy to make this happen, the predictable debris left over from such a process would seem to support the previous explanation of flying rocks being responsible for creating craters. However, I would ask you to consider that these two explanations have become linked through a form of lazy convenience, and I do not mean a deliberate act. It is a fact, however, that if this convenient association between cratering and planet formation were disrupted, it would be another very unwelcome turn of events for supporters of the gravity model. Well, this is what we are about to do, because the majority of craters have nothing at all to do with flying rocks. Moreover, we will see later that planets and moons were not born from vast collections of pilotless, relatively slow moving pieces of rubble.

But wait; is it not obvious that flying rocks would form craters and that this theory simply makes sense? After all, most of us will have thrown stones into mud and sand and can recall the crater-like results. This simplistic idea of the effect of an impact might therefore persuade us that this is a subject that needs no further investigation. Well, I can assure you that there is indeed another line to consider. It is one backed up convincingly by evidence from electrical and plasma laboratory experiments; a much better situation to be sure, in comparison to what we find on the table from scientists convinced about impact theory. The electrical explanation for craters and other surface markings will be provided in detail later, but for now, I will say that the same basic reasoning for nearly all crater production, no matter where they are found, is at its core.

Along the same lines, questions are raised about other types of surface markings and geological features. I will also explain the likely causes of these, where we will see that the same basic electrical processes can be shown to be responsible for them all. It is relevant here to note that just a few fundamental electrical theories will be at the heart of everything I describe. The things that have turned out to be problematic for today's astro-science all have explanations within the Electric Universe model.

## Other relevant things ...

Here are a few more examples that also have questionable aspects to them. These serve in addition, to highlight the waste of resources that has gone on and still goes on; resources which otherwise could be applied to far more constructive research and concept development. Some of these examples are quite involved so I will not be going deeply into them; I do not have the knowledge or experience for that. I therefore present them in summary form as the work and thinking of current professionals in these areas. I want these examples to provide an appreciation of the issues they refer to and to show they are relevant to mention here.

... Einstein told us years ago that there are three physical dimensions; up and down, left and right, and towards and away from us, plus one other, that of 'time'. He came up with a label for this which he thought would describe the space around us and everything else, everywhere. He chose to call this label 'space-time'. The idea was that we, or any object, could move in any physical direction in space, (in outer space or here on the ground) but that any actual movement also involved the passage of time. This is why he included it as the fourth dimension. In the three obvious physical dimensions of outer space, this also described a region of gradually changing gravitational force around all bodies that would act as if it were a hole in space that pulled things in. This concept was seen as 'leading edge' and fitted well with associated thinking, so the term stuck immediately and it has been used seriously ever since. Space-time has for a long time been awarded the status of 'a real aspect of our reality'. It is a concept that scientists have come to view as correct enough to include within their own work. Many experiments have been conducted to prove it is real, just as Einstein said it is, but in fact, the concept of space-time seems to have no real meaning, so what have the experiments to prove it achieved?

One such very expensive experiment was set up to detect so-called 'Frame Dragging'. This is what is said to be happening as the Earth rotates in its own 'gravity hole' from where it supposedly 'drags' surrounding space-time around with it. You will not be surprised to learn that this kind of experiment cannot be done cheaply or easily, and one of the options pursued has indeed been to carry an experiment out into space. At the start of that project, money was spent on existing space probe missions in an attempt to see if their data or data archives could provide proof of frame dragging - this came up with nothing. Later, funding for a more tailored attempt was provided in 2004 for the space probe mission "Gravity Probe B" [3-4]. Still, no evidence of frame dragging was found. A substantial amount of money had therefore been spent on the project with no result, with the outcome being that the fundamental theories upon which it is built have taken a very serious knock. This situation reflects badly on Einstein's ideas and the extent to which some scientists have been determined (and allowed) to pursue work that seems now to have had very shaky foundations in the first place. One might say "it's okay to say this in hindsight" and I would say in response, this is not the way to look at it. This is because we have an example here of project work founded on known questionable theories as the basis on which to seek valuable funding. It is obvious, however, that these project proposals have been presented so cleverly to the purse holders in astro-science, that they henceforth allowed them to become exceptionally well-funded projects. This is an example of building a telescope to see something that only exists in the minds of theoreticians.

... A ground-based experiment known as LIGO [3-5] (Laser Interferometer Gravitational-wave Observatory) was set up in 1992 to detect what are known as 'gravity waves'. These waves are thought to be 'ripples' in the fabric of space-time, and the claim is that they should be detectable. Over many years since the LIGO project started, major experiments have been run many times at various locations across the world and hundreds of scientists from high profile scientific institutions have been involved. Their sensing equipment has often been left in its detection mode for more than complete years at a time ... and nothing has ever been found. Nevertheless, they remain convinced of their theories and continue to be supported very nicely by public funds.

There has also in the past been a great deal of work done and money spent on a spacecraft mission called LISA (Laser Interferometer Space Antenna) – at least, this was up until the project's recent expensive cancellation. The 3 probes involved in this mission would have flown in strict formation with 5 million kilometres between them in another attempt to detect the gravity waves that the LIGO experiments have failed for almost 20 years to find. One does wonder what stories are manufactured by financially astute project heads to sell their ideas to the decision makers. The science at the root of this gravity wave issue is being questioned, and again, this reflects badly on Einstein's theories and no alternative ideas are being considered.

... In relation to the dark matter I mentioned earlier, an experiment known as CDMS [3-6] (Cryogenic Dark Matter Search) was set up in 2002 to prove the existence of that illusive commodity by searching for the particle it was thought at the time to consist of - this being called a 'WIMP' (Weakly Interacting Massive Particle.) Again, nothing has been found and lots of money and resources have apparently gone down the drain.

... I previously mentioned the Cosmic Microwave Background (CMB) which our instruments indicated was evidence of a very smooth 'energy/temperature background' to the universe. Well, despite the fact that this discovery was conveniently taken to support the idea of 'inflation', the recent WMAP [3-7] (Wilkinson Microwave Anisotropy Probe) space mission has through its higher resolution sensors detected a significantly uneven spread of hot and cold regions across the CMB. This spread of uneven temperatures indicates that things out there are not as smooth as the experts originally thought they were. In actuality, the very existence of this more uneven temperature profile contradicts the idea of inflation and other predictions derived from Einstein's theory of relativity.

... If we consider again the idea of an expanding universe with its far-away galaxies supposedly travelling further away and gaining speed all the time, then these galaxies should be dimmer than the galaxies that we know are close to us. I think this sounds reasonable. It turns out, however, that far away galaxies are of similar relative brightness to those that are close to us. An example of this is galaxy HUDF-JD2 [3-8] Again, the implication here stands in direct opposition to the idea of an expanding universe.

... The galaxy M87 [3-9] (Virgo A) is seen to be ejecting material at faster than light speed. This again contradicts Einstein's theories.

... There is a cluster of galaxies 600 million light years away from us referred to as Abell 3376 [3-10]. In 2006, this cluster was observed to be surrounded by a 6 million light year in diameter ring structure of glowing material, the temperature of which was measured to be in the region of 60 million degrees Kelvin. This is many tens of times hotter than the hottest area associated with our Sun. Again, there is not a snowball's chance that the 'mechanical shock' theory of the gravity model could ever explain this extreme level of temperature generation. It is also interesting to note that NASA pages on the web do not refer to or even hint at this issue. Once again, the gravity model and its proponents fall short in their ability to address what we see in the real world.

... There is an Australian physicist by the name of Stephen Crothers [3-11]. I mention him because he has put forward what seems to be proof for many scientists that Einstein's equations that led to the subsequent invention of Black Holes, are fundamentally wrong - the interpretation of the mathematics appears to be wrong, believe it or not! In addition to this being a challenge for many to accept, it should be kept in mind that Einstein himself became an objector to the Black Hole theory. You should also keep in mind that it was the use by others of Einstein's equations that established the theory of Black Holes. Despite all this, his name remains firmly linked with the idea that they are real. It is odd that nothing is ever done to clarify any of this. I wonder, therefore, if conveniently doing nothing supplies ongoing credibility to the idea of Black Holes. For anyone looking for more detail on Black Holes, I would refer them to the work of retired Professor of Physics and Mathematics Jeremy Dunning-Davies of Hull University and his book "Exploding a Myth" ISBN: 978-1-904275-30-5 [3-11A] [3-11B]. In this book, Jeremy, as an expert commentator on the history of science, supplies a detailed but easily digestible description of the truth behind the complete history of Black Holes.

… Young stars have been found in locations where according to standard theory they should not be [3-13] – this being at the centres of galaxies. The electric universe model yet again can supply an explanation for this.

… As previously mentioned briefly, the space probes known as Pioneer 10 and 11 have been observed not to be doing what they should be doing out at the very edge of our solar system right now [3-14]. They were launched in 1972 and 1973 respectively, where their initial missions were to inspect Jupiter and Saturn. However, they were subsequently re-tasked to gather data and images as they travelled on the rest of their journey to the edge of our solar system and beyond. They are both now at the boundary between our solar system and deep space, and the situation is that they are both substantially off course, slowing down and covering 3,000 miles less than they should each year. Something has deflected them and whatever it might be that is holding them back noticeably, it is certainly not the Sun's gravity! Answers to this have been pursued for a while now but no explanation from standard theory can be found. Due to the evidence gathered and in what is judged by many as a previously unthinkable admission, some astro-scientists have been forced to consider the possibility that they might not fully understand the laws of physics! Well shoot my boots off! … I think they should read about the Electric Universe model to get some pointers to reasons for what they are seeing.

… Saturn's polar regions have been found to emit radio waves. [3-15]

… Gravity and mass can vary; they appear not to be constant and it seems they both have a relationship with the electromagnetic force. [3-16]

… The geological record shows us that gravity must have been lower in the time of the dinosaurs than it is today for those creatures to have grown so big. If this had not been the case, then it is calculated that their bones and muscles could never have supported the weight of their bodies as we now know them to have been. [3-17]

---

The goal of this chapter has been to present some of the main questions that exist around the accuracy and acceptability of the gravity model of our universe and to give the assurance that answers and better clues are available from the Electric Universe viewpoint. None of these are minor questions and all of them are out there and available to inspect in detail by anyone interested in making their own investigations. No doubt you will have detected in the words I have used, my personal disappointment, frustration and at times, a lack of respect for some of those who inhabit today's astro-science establishment. I make no apology for this.

In this piece of the journey to explain the basic EU model, we have laid out the case it stands in answer to. The next chapter will provide an opportunity to learn about, or to revise, the basic scientific, electrical and plasma theories that will help in gaining an understanding of what I go on to describe in chapter six. You could skip over this next chapter if you feel you do not need to cover the basics, but there are a few EU things in there as well.

# 4 | Some basic theory that will help ...

This chapter covers basic technical and scientific information that will be helpful for understanding the theories that underpin the Electric Universe model we look at in chapter six.

### What is an Atom?

Everything we are aware of is made up either of atoms or their component parts. In terms of Earth, this means the air we breathe, the water we swim in and the ground we walk on. We believe it applies just the same to the make-up of every other body or physically detectable presence in the universe as well. Anything you can possibly think of consists of these incredibly tiny structures. The component parts of atoms are much smaller structures referred to as sub-atomic particles, and in terms of the basic atomic model, the names we give them are Electrons, Protons and Neutrons. All of these particles are electric in nature. An electron has a negative (–) charge, a proton has a positive (+) charge, and a neutron has equal negative and positive charges, so we say it has no charge because those charges balance each other out to give an overall neutral charge.

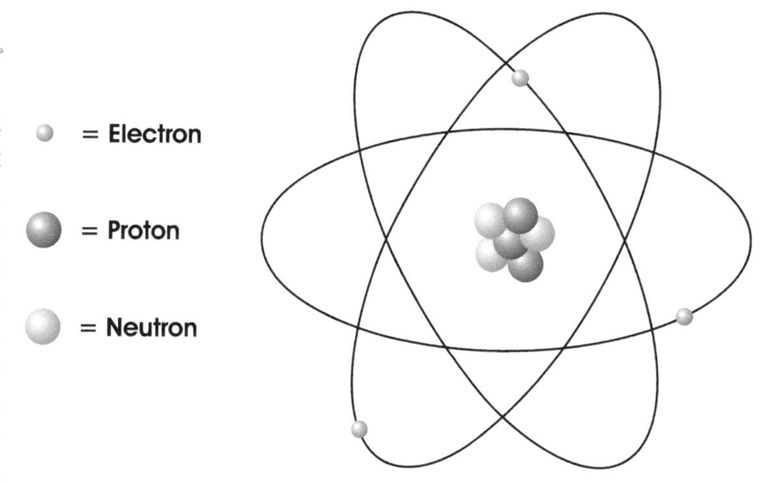


The basic model of an Atom © author ▷

The structure of an atom can be described as follows: At its centre it has a nucleus that consists of one or more sub-atomic particles. The nucleus is where we find all the protons and neutrons. Electrons, of which there can also be one or many more, whizz around this nucleus in 'shells' at set distances from it. The reason for these shells having specific distances from the nucleus is that the electrons they accommodate are grouped by distinctly separate levels of energy, and no other levels in-between. Gaps therefore exist between the shells. When bound to an atom, electrons are found in shells that best suit (or are best 'tuned to') their own level of vibrating energy. These concentric shells are also sometimes referred to as 'orbits'. In the case of an atom with many electrons, the smallest shell closest to the nucleus is where the electrons vibrate least rapidly and have least energy - the largest shell is the one furthest away from the nucleus where electrons vibrate the most, and therefore have the most amount of energy.

In the normal state where everything is in balance, the sum of the charge of the total number of positive protons in the nucleus balances out with the sum of the charge of the total number of negative electrons in their respective shells. This state of balance defines an overall electrically neutral atom. Here we remember that the neutrally (or zero) charged neutrons that are normally present in the nucleus along with protons, have no influence on the overall charge state of the atom.

The simplest atom we can consider out of all the available elements is the hydrogen atom. It has one proton as its nucleus, one electron orbiting in a single shell, and no neutrons.

The simplest Atom - Hydrogen © author

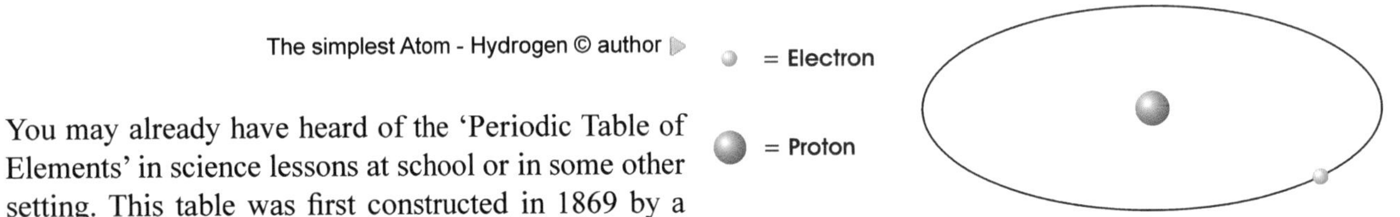


You may already have heard of the 'Periodic Table of Elements' in science lessons at school or in some other setting. This table was first constructed in 1869 by a Russian chemist called Dimitri Mendeleev. Its contents have grown since then and it now lists 92 naturally occurring elements and 26 that have been created by us, so that is 118 in all. Elements such as hydrogen, iron, oxygen, copper, carbon, sodium and tin are all basic types found in nature, and when elements combine we find 'molecules' formed of other things such as water, carbon dioxide, salt and glucose. It is the single elements of the periodic table and combinations of them as molecules that give us all the things we are currently familiar with in our material world. For instance, molecules of water are made up of the basic elements, oxygen and hydrogen, and dry air is made up mainly of molecules of nitrogen, oxygen, argon and carbon dioxide.

The sole difference between these 118 elements comes down to the numbers of protons, neutrons and electrons that each of them have. Here, you would be correct in assuming that all these sub-atomic particles are the same; a proton is a proton, a neutron is a neutron and an electron is an electron - it is just the numbers of these and energies involved that make the differences between all the elements. We have already said that a basic atom of hydrogen gas is made up of one proton and one electron. This is as basic as we can go with atoms themselves, so hydrogen is considered to be the 'lightest' element in terms of its mass. In contrast, an atom of copper has 29 protons, 29 neutrons and 29 electrons, and is therefore a much heavier atom than hydrogen. This makes sense because we know from our daily experience that metals are heavier than gases. This is important to point out because you can think of it as the reason why all different types of matter have different weights. Materials like aluminium and lead are both metals, but in the same physical volume of, say, one cubic centimetre, their weight is significantly different for precisely this reason. If you spend a moment, I am sure you could come up with your own examples.

I said that copper atoms have 29 electrons and that these are found in separated shells around the nucleus.

The copper atom © author

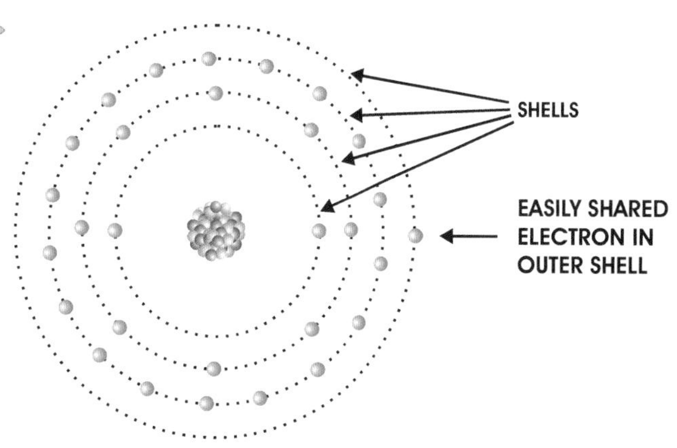


Copper atoms are very good at sharing the electrons in their outermost shells with those of other copper atoms close to them. Most other metals do this too but copper is a good example to consider in our book's context because it is the material we are most familiar with when we consider the type of electric conductors we find inside the cables carrying electric current in our homes. This electron sharing activity provides a 'gluing effect' that binds the atoms together into a kind of lattice structure in which they all share their outer electrons with those of their neighbours to hold that lattice together. However, in addition to this natural bonding, the outermost 'free electrons' can be made to move away from their atoms to other atoms by the application of an influential force. This force can be either internal or external to the atomic structure, which itself can be a typical electric wire. The result is the same; all free electrons will be influenced to move in the same direction at the same time. This common flow of electrons (sub-atomic particles) is what we call an electric current. The atoms of some materials, mostly non-metals, hold on to their outermost electrons very tightly indeed; they are not very keen to lose or exchange electrons with other atoms. In fact, for all practical purposes, some of these atoms just refuse to allow the sharing of electrons at all. The materials these atoms make up are mostly non-metallic and are themselves, or in combination with other materials, known in electrical terms as 'insulators'. These materials do not allow electric current to flow through them easily; examples being glass, plastic, rubber, ceramic, air and wood. This is why our domestic electric cables normally have their central copper wires safely covered in plastic or rubber. If we fragile humans are to enjoy long and happy lives, then we need to keep those powerful electric currents away from our bare skin by shielding them in this way.

A few other things that should be pointed out about atoms are; the relative mass of their protons, neutrons and electrons; the way in which these are distributed inside the atom and what happens when electrons change from one shell to another. The mass of a proton is around 2,000 times greater than that of a single electron. One neutron is the same mass as one proton plus one electron, so it is the heaviest of the three sub-atomic particles we have so far mentioned. If an electron is persuaded to move at a particular speed to strike some other particle, it will not make as big an impact on its target as a proton would at the same speed, and a proton would not make as much of an impact as a neutron would. The fact that these different sub-atomic particles have different masses is an important aspect to keep in mind for later.

We will go back to our model of a simple hydrogen atom with one proton and one electron to consider 'scale'. Think about the single proton (the nucleus in the centre) as being the size of a normal soccer ball. Think then about the electron being the size of a pea. If you were to stand with your foot on the soccer ball, then in terms of the relative distance between the nucleus and the electron in a real hydrogen atom, the pea would be around 25 miles away from you! Everything else in between would be empty space. You can see from this that atoms are nearly all made up of empty space, including, of course, those of which your body is made! Things are this way because an exceptionally strong electric force exists that allows atoms themselves and structures consisting of them to bind together. We shall be looking closely at this force along with the other forces we know about later. A sobering thought comes to mind here due to the fact that all atoms have so much empty space inside them. This is that everything you can see, touch or even think about, is mostly just empty space! If a low number of electrons are made to move communally and slowly, then not much energy will be involved in that process, but if many are made to move communally and quickly, then there will be a lot of energy involved. This helps us understand how 'current flow' in a conductor can vary in value. A small current value is when a low number of electrons pass a certain point in a given amount of time, and a large current value is when a high number of electrons pass a certain point in the same amount of time.

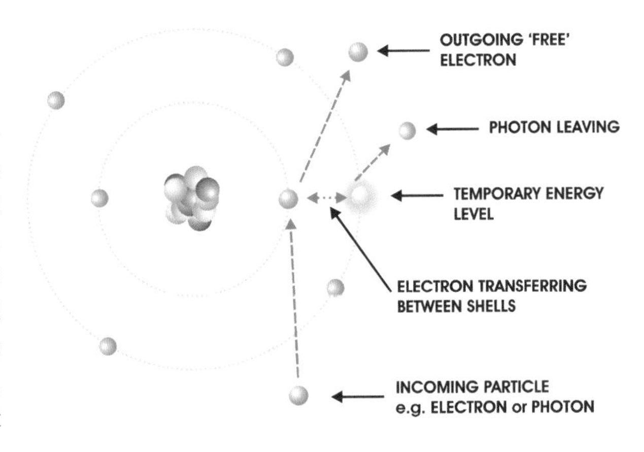


How an atom emits a photon of light © author

If an electron is struck by another fast moving (high-energy) particle, for example another electron, then the energy it absorbs from the incoming particle can move it into a higher energy shell that is further from the nucleus. The energy balance of the atom must be restored, so the electron then returns to its original lower-energy shell, and by doing so releases a 'packet of energy' that represents the energy difference between the two shells involved. This energy packet is a Photon, an uncharged particle that can then go on to cause the same effect with other atoms.

A photon's level of vibration and energy is determined by the shell that its emitting electron came from (the outer one). We humans mainly detect photon energy as light by our eyes and as heat by our skin. Photons can also carry very much lower and very much higher levels of energy; conditions which we shall discuss later. Consider this example: we detect photons of light emitted by the Sun as our impression of daylight scenery in our normal environment. When these photons interact with and are re-emitted from common objects such as clouds, trees, mountains and grass, the light receptors in our eyes discriminate between the different levels of energy vibrations imposed on the photons by the atoms in the materials from which they came.

We identify these different vibrating energy levels as individual colours and intensities of light. These in turn are what form the images detected at the back of our eyes by our retinas and sent as electrical messages to our brain. Our brain then determines all aspects of what we are looking at so that we can make 'visual sense' of the world around us. This is how we see and understand objects in our environment, a process which further allows us to make decisions on how to interact with the reality of which we are visually conscious. The same general process applies to the bio-electrical signals received by our brains from all our other senses as well. This, however, is a separate and very big subject area, so we will stop here and return to considering atomic structure. The important thing is that electrons, when forced by the application of some form of external energy, can briefly change the shell they are in and emit a photon of light that has an energy level directly associated with the properties of the element that the photon came from. Energetic photons are what we detect as heat and light of different colours, plus other forms of radiation that we will touch on shortly. At this point, I will expand a little on my reference to 'vibrating energy'.

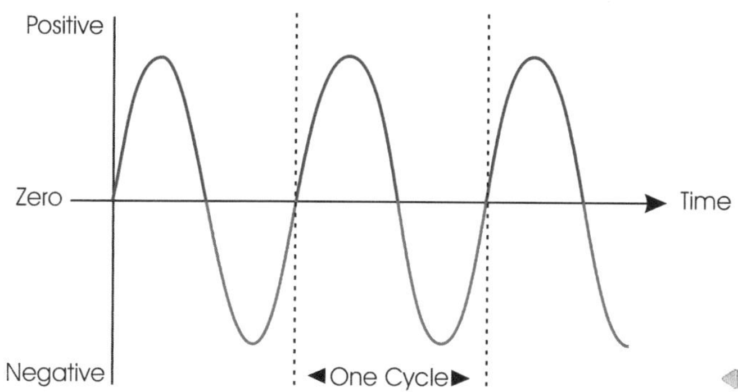


What I intend by my use of the word 'vibrating' is to introduce the idea of 'frequency' or 'oscillation', as it is sometimes referred to. Both these terms are very common, very important, and closely related. This is where a physical object or some force, changes (or cycles) back and forth between one value and another, repeatedly (alternately) and usually at a steady rate.

The idea of one 'cycle' © author

The energy level of mains electricity in our homes does this at either 50 or 60 times per second. This is considered to be a very low frequency in the grand scale of frequencies that exist in the universe, for as we will see when we cover the idea of the Electro-Magnetic Spectrum, frequencies go to very much higher levels indeed. We should clearly establish a couple of things here:

- In a stable situation, electrons have a particular frequency associated with them that represents the energy level of the shell in which they orbit their nucleus.
- When electrons are forced to move between one shell and another, and depending whether the move is to a lower or higher energy shell, they emit or absorb a photon that has a frequency/energy relationship determined by the difference between the shells involved.

There is much more that could be said about atoms, how they behave, how they are influenced and how they can be combined, but to suit the level of this book, this amount of information is enough.

## The four fundamental states of matter

We are all familiar with solids, liquids and gases, but have we ever heard of these with the term 'plasma' mentioned as well? Plasma belongs here as it is also a form of matter. Altogether, these four states of matter are associated with every form of physical thing we can see or think of in our world and in our universe.

Our four states of matter are (1) plasma (2) gas (3) liquid (4) solid.

To help with drawing mental pictures, I will base the following explanations around something we are all familiar with by referring throughout to the various forms in which find water and to its constituent elements. Water is a molecule $H_2O$ that has two atoms of hydrogen and one atom of oxygen.

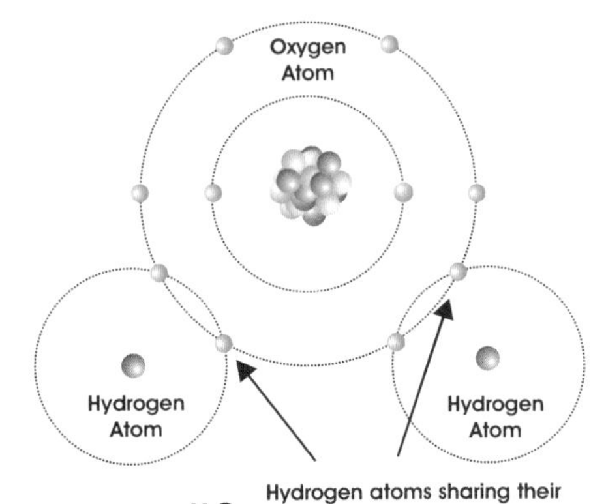


© author

**'Solid'** is the fourth state of matter. Water ice, just like iron or diamond, is considered a solid because its molecules are aligned rigidly in a crystalline structure. Ice forms at 0˚Celsius (Centigrade) because the heat energy contained in the atoms of hydrogen and oxygen that form its molecules is very low. In other words, the electrons of the atoms that make up the water molecules are not oscillating with enough energy to cause breakage to the bonds they form between the molecules of the crystalline structure, so everything is held together to produce solid ice.

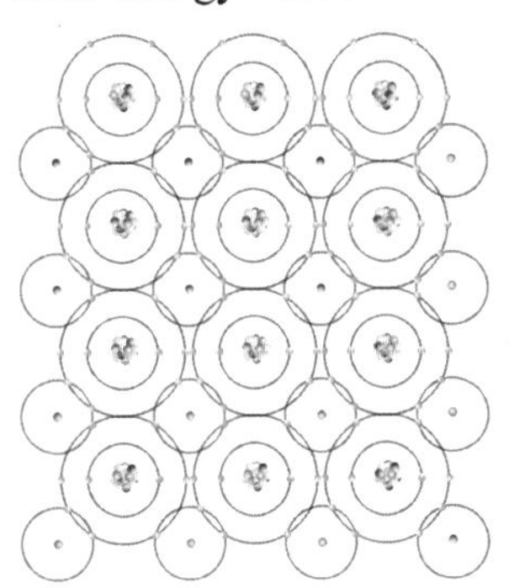

Lattice structure of Ice © author

**'Liquid'** is the third state of matter. When the temperature of ice rises to just above 0˚C it begins to melt and become liquid water. This process occurs because the heat energy contained in the atoms has increased enough for electrons to begin breaking the bonds between the molecules of the crystalline structure of the ice. In other words, the energetic oscillations of the electrons of the hydrogen and oxygen atoms of the water ice molecules have been increased to a great enough level through the application of external energy, that separation takes place between the inter-molecule bonds. This can also be seen as an overall increase in energy stored within the ice structure. A general point to make here is that because the molecules are less restricted, liquids are considered to be 'less mechanically stable' than solids.

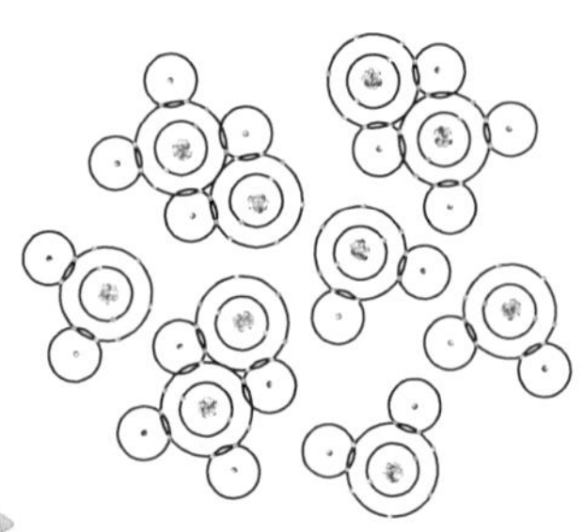

Unstructured water molecules © author

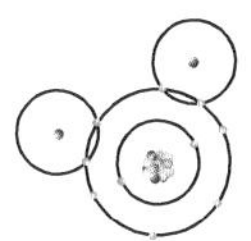

**'Gas'** is the second state of matter. As the temperature of liquid water rises, some of its molecules will separate off and rise away into the air as water vapour gas.

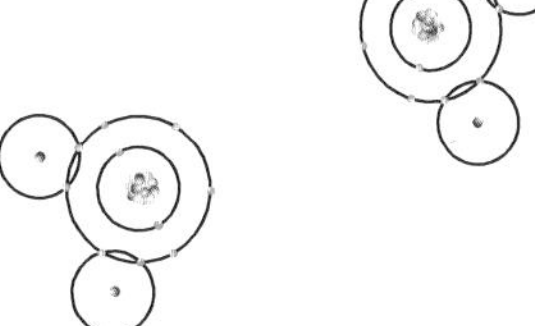

Sparse molecules of water vapour © author

With no deliberate application of energy, this is what we would consider to be the normal evaporation of water as it turns into a gas that is lighter than air then drifts away. We get precisely the same result but much more of it when we deliberately heat water up so that it evaporates rapidly, like when we raise its temperature to boiling point at sea level of 100°C and it turns into steam. During this process, the molecules give up their relationships with each other to become a thinly dispersed molecular gas where the individual molecules float about in a random fashion. Beyond this process we can, through special means, separate the individual water molecules into their oxygen and hydrogen component gases.

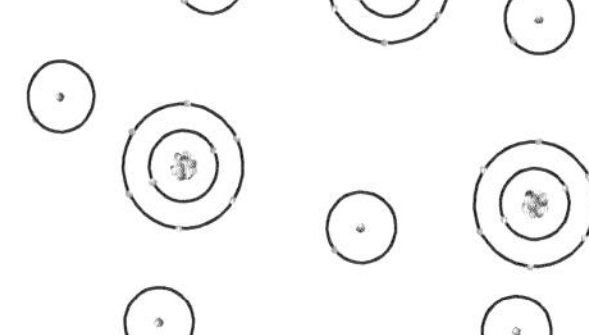

Atoms of hydrogen and oxygen separated © author

For the level of understanding we want to achieve, this process of increasing or reducing the energy that electrons have in order to bring about the various gas, liquid and solid states, can be considered as the same process that applies to all types of matter. (Note: Some solid combinations of matter take shortcuts and do not appear to work like this - an obvious example is wood - you cannot heat wood and produce liquid before a gas, it breaks down and goes initially to other solid matter and gases.) For our overall discussion, however, it is useful to accept for now that at large structure levels, where we have combinations of molecules and elements that make up all physical things in our environment, the basic conversion rules as described here will apply. Here is another example ... Think about steel turning to a liquid form when heated to its melting point then separating into atoms of the gases Iron, Carbon, Manganese, Phosphorous, Sulphur and Silicon as the temperature is increased even further. In a broad sense, when considering matter, it all seems to come down to various amounts of energy being present at the molecular and elemental bonding levels that gives us the range of materials in our world with which we are familiar - the air we breathe, the water we swim in, and the land we walk on - and every other physical thing you can think of as well.

Most gases, liquids and solids share an additionally important property, this being that they have electrons which to some extent can freely move about or be easily coaxed into doing so. If an appropriate force is applied collectively to these free electrons to make them move in the same direction, then this can be considered as producing (or 'inducing' being the more correct word to use) an electric current to flow in that particular form of matter. The point here is that all forms of matter can theoretically conduct an electric current; it just depends on how much of an 'encouraging force' is applied to the electrons in its atomic structure to make them move. Note that this will apply not just to the atoms of the 118 elements but also to combinations of these in all the varieties of molecules these elements form. Here you can consider the fact that tap water is a good example of an electrical conductor, and this is why you should never play with mains connected electrical gadgets while you are taking a bath! Now we can move on to look more closely at **Plasma … the first state of matter**.

**Plasma**: what exactly is it? Well, if you continue to add energy to the atoms in a gas, eventually, some of the outer electrons will be stripped off the atoms to become free electrons. The atoms left behind will therefore have a net positive charge. The result is a gas that can conduct electricity and respond to electromagnetic fields. In line with the collective opinion of far better qualified folk, I consider plasma to be 'the first form' of matter. Some people would dispute this definition but I think it is substantial enough to defend, because from plasma, everything else can follow. It is interesting to note that supporters of the Big Bang theory often talk about there being nothing but plasma in the beginning!

An active plasma state exists within any form of matter that has an electric current flowing through it. Unlike neutral matter that is made up of electrically balanced molecules and atoms that can be influenced by gravity, the actual active plasma (current flow) within any form of matter will not be influenced by gravity. Do you hear of cases where it is critical to have electric cables positioned in such a way that ensures the flow of electric current is in a downward direction into an electrical device? No, you do not. This is because there is an electrical pressure force that we call 'voltage' which pushes the electrons along the conductors of the cable. This is the same in every situation that involves matter in the plasma state. Gravity has no effect on electric current flow because the electric (electromagnetic - EM) force is far stronger than the force of gravity. Note here that the normal matter we interact with in our everyday lives on the surface of the Earth is mostly in an electrically balanced state and is therefore not influenced by the EM forces in our environment. Electrically neutral matter will therefore only respond to the force of gravity until an electrical imbalance in that matter is brought about; then the EM force can have an effect as well. In our everyday electrically neutral lives, we have no real direct contact with or awareness of electric current flow – unless, of course, we touch a bare live wire!

Plasma is based on EM relationships, where electric fields and electric currents, with their associated magnetic fields, are the powerful influences that dictate what matter does when it is in the plasma state. We will later describe in greater detail what electromagnetism is, but for now, I will provide you with the most potent fact about it. The electric force is one thousand, billion, billion, billion, billion times more powerful than the force of gravity; this is the number 'one' with thirty-nine zeros following it, or to write this out in numeric form we have 1,000,000,000,000,000,000,000,000,000,000,000,000,000 … or $10^{39}$ in scientific notation terms.

This fact really drives the point home about which force can have the greatest influence on forms of matter ***when it is in an electrically imbalanced and therefore plasma state***.

This also brings one to wonder about a couple of other things; why this well-known fact is not more widely included in astro-science discussions and why, with their undoubted awareness of the EM force, the focus of the establishment's research has traditionally always been on the force of gravity. Here are some everyday earth-bound examples of where the plasma state of matter exists.

**Gas**. The gas that glows inside a common neon tube is in a plasma state. It appears to us this way because an electric current is being forced to flow through it, a process that causes electrons to collide with atoms and light (radiation) as visible photon energy to be emitted.

**Liquid**. The dilute liquid acid solution inside a lead-acid car battery allows the passage of electric current through it while it is being charged from an external source. In the opposite direction, it also allows a current to pass through it when the battery supplies its stored energy to, for example, the electric motor that starts the engine of the car. In addition, it also allows the passage of a very small amount of current through it while the battery is not in use. This natural action tends to discharge the battery and is called a leakage current. The current-carrying liquid in a lead-acid battery is known as the 'electrolyte' - it is liquid matter in a plasma state.

**Solid**. Here we can legitimately consider the copper wire inside an electric cable that has an electric current flowing through it. Although the lattice structure of the metal atoms of the conductor is very rigid and typically remains that way, the flow of current within the copper wire defines *that* solid metal structure as being in a plasma state.

These brief examples serve to tie all forms of matter together with the plasma state, but there are other very important facts to give. Current-carrying plasma presents itself to us in three states – these so-called 'plasma modes' being Dark, Glow and Arc, where each is considered to be a distinct state.

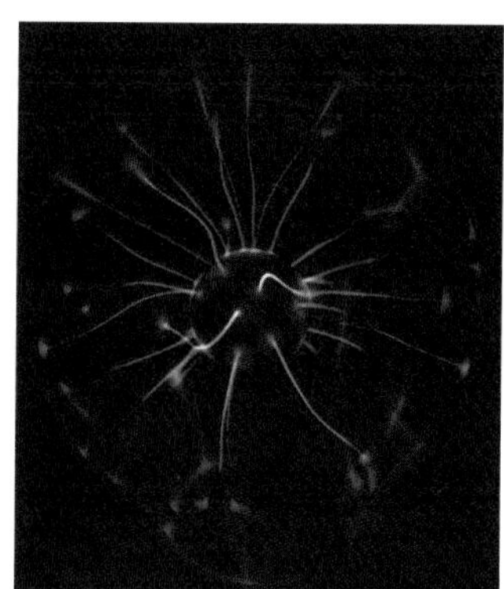

A decorative Plasma Ball © author ▷

**Dark Mode** is where we cannot see the plasma current but it is definitely there and it is definitely doing something at a relatively low power level.

**Glow Mode** is where we can see what is going on with the plasma because photon energy is being given off as visible light due to the presence of a greater amount of energy and the subsequent energetic behaviour of sub-atomic particles.

**Arc Mode** is matter in an extremely energetic (hot) plasma state, the radiation from which can be scary on large scales and severely damaging to humans!

I will give some examples of where these modes are apparent and point out that the only difference between them is a combination of the electric pressure (voltage) applied and the density of the current available to flow within the plasma.

Dark Mode: (Low Energy)

- A household 'Ioniser' for air purification.
- The flow of electric current during the 'Electro-Plating' process.
- The beam of electrons that flows inside the Cathode Ray Tube of a television.
- The flow of electric current in your body that you feel as a shock when you touch a live wire.
- The flow of electric current through the Liquid Crystal Display of a computer screen.

Glow Mode: (Medium Energy)

- The action inside a neon tube when its gas is stimulated to give off light.
- The glowing filaments you see inside a decorative Plasma Ball.
- The glow that results on the screen of a Cathode Ray Tube when it displays visible images.
- The glow given off from a Comet's tail.
- The Aurora we see in the sky over the North and South Poles.

Arc Mode: (High Energy)

- The Lightning we see during an Electric Storm.
- The blinding glow we should not look at when Electric Arc Welding is being done.
- Action in the Sun's Photosphere and on other stars as they give off brilliant light and heat.
- The brightly glowing areas that erupt on Comet surfaces as they get close to the Sun.
- The spark that is generated by the Spark-Plugs in your car engine to ignite the fuel.

These are all examples of plasma with different levels of voltage pressure applied and different levels of current density flowing through it. We cannot 'see' electric current flow, we can only see the effects of it. If there is a large amount of energy involved then we need to take care because the photon energy given off as radiation can be significantly dangerous for us. A good example of this is with electric arc welding. Here, high-energy photons are radiated away because powerful electric currents flow that melt metal. This EM radiation is in the form of ultra-violet rays and X-rays, both of which can seriously damage our eyes and other body tissues.

This is why people who weld wear visors that have special glass to stop harmful radiation from passing through. The same goes for when we may sometimes be tempted to look directly at the Sun; do not do it, for there is a significant amount of ultra-violet radiation present. Although the Sun's perceived brightness is itself damaging, it is the invisible radiation that will do us most harm.

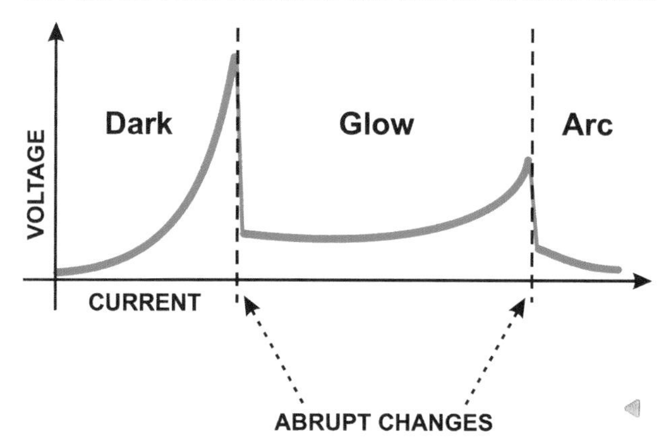


Here is a diagram that shows us the differences between the three states of plasma. Note the relationship between voltage and current levels. As the voltage increases while a large current is available to flow, the plasma will transition very abruptly between each mode. You may recall what happens when a fluorescent light is switched on. In this diagram, the sharp transition between the dark and glow modes is what you see as a short flicker before the tube reaches a steady output in the glow mode.

The three plasma states © author

We said previously that neutral atoms are that way because they have a balance of positive and negative charges between protons and electrons and that their neutrons play no part. With this in mind, here is a more precise description of what plasma is.

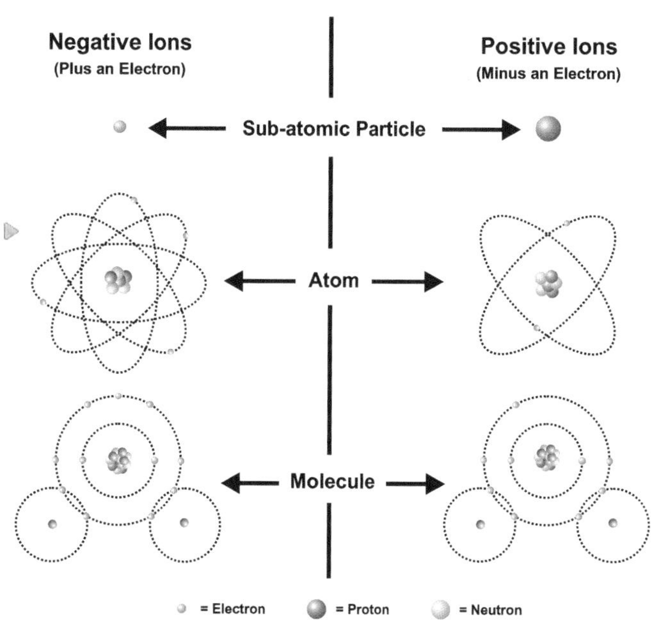


Electrons, Ionised Atoms and Molecules © author

If a neutral atom were to lose one of its electrons it would be defined as a 'Positive Ion' because it would be left with an overall positive charge. The electron then free to move around could be called a 'Negative Ion' because of its overall negative charge. So here we see that it is the absence of electrons that defines positive ions, and electrons on their own or in groups can be defined as negative ions. Take note that just as individual electrons are defined as negative ions, individual protons are defined as positive ions.

If a hydrogen atom has its single electron and single proton separated, then one negative free electron and one positive ion will be the result. Hydrogen will be very important to consider as we progress, so remember this in particular. Additionally, molecules of gas and solid matter such as dust that have an overall positive or a negative charge because of either an electron excess or deficiency, are also included here in the definition of ions. Any atom or molecule needs only to have one single electron missing or in addition to be defined as such. Be aware also that the process through which ions are formed is called 'ionisation'.

The plasma state is found in gases, liquids and solids everywhere. It is therefore not wrong to consider that with the exception of similar conditions on other planets like our own, absolutely everything outside our thin 'electrically deceptive' protective biosphere here on the surface of Earth, is plasma in one or other of its states. This is why we say the universe is made up of 99.99% plasma.

**Temperature and energy:** Matter that is not being unduly influenced by external electrical or magnetic forces may tend to have a natural 'drift' of electrons and charged molecules within it. This minor activity manifests as a constant 'balancing out' interaction of ions that can be viewed as the natural 'background exchange of energy' that produces an overall electric neutrality for the matter in question. This particle motion is random and is brought about by energy (temperature) in the environment. It is known as 'Brownian Motion'. Think of it as the natural level of energy that something has before its temperature is changed by an external influence. It is also common to refer to the 'temperature' of matter as its 'energy level' or 'energy content'.

As an example, consider where we have air in a sealed room and no special conditions in place. The molecules of oxygen, nitrogen and other gases that make up the air will float about in a chaotic mix, occasionally bumping into one another and perhaps exchanging the odd electron here and there. This low level of energetic motion represents the 'natural temperature' of the air in that room. The energy involved is not very useful while it is at normal room temperature. However, it would become useful if the temperature of the air were raised through some means to a higher level, especially on a cold day. In that case, the molecules of air would be moving around in a faster, more chaotic fashion so the energy they would be able to impart to other forms of matter might be felt, for instance, by us as heat on our skin.

At the low end of the scale of temperature we have levels of coldness. If we could cool matter down enough so that all sub-atomic motion is stopped, then that matter would be said to have no energy associated with it at all. This, however, is an apparently impossible level to reach, for everything we are aware of must maintain just the tiniest amount of energy for its structure to be preserved. This level of zero energy is theoretical and has a temperature scale associated with it that starts at the 'no energy level of zero'. This is known as the Kelvin scale where zero degrees Kelvin equals no energy whatsoever, no movement of electrons, molecules, or anything. Nothing can reach 0K, but science has come very close indeed. In terms of how the Kelvin scale relates to the Celsius scale, 0˚C is the same as 273K, so this means that minus 273˚C is the same as 0K, or what is known as 'absolute zero'. Note that temperatures quoted in Kelvin do not usually have the degree ˚ symbol included.

## Electricity

This should be easier for us to understand now that we have looked at the basic building blocks. Electric current flow results when an electric or magnetic field is applied to matter in a way that produces within that matter, a communal flow of electrons. As a concept, this will be the case with any form of matter. If we put magnetic fields to one side for now, the force to consider first is the electric field. This is the force that is responsible for us receiving a shock from a door handle when we walk on a nylon carpet and the one that allows us to stick balloons to walls by first rubbing them on something made of wool or nylon. It is called the Electro Motive Force ('emf' for short). The effect of this force is to align the negatively charged electrons inside any form of conductor like strings of connected train carriages. If an electric pressure force is applied at one end of this string, the effect is seen instantaneously at the other end. If a complete circuit is then formed, the resulting movement of these communally aligned electrons past a given point in that circuit is what we call electric current. In truth, the flow itself is excruciatingly slow by our normal understanding of speed. I will also point out that the word 'electricity' is not a very precise term because it does not just represent one thing; it actually includes two components:

The first of these is the voltage or emf force that influences electrons to move communally in a particular direction. This is the force we sometimes call 'potential' or 'pressure', or if we are comparing it to another level of voltage, the 'potential difference'. All these terms really mean the same thing and the unit used to measure voltage is the 'Volt'. The second component is the density of the flow of electrons when a voltage difference exists to make that happen. This flow, referred to as current flow or just current, is measured in 'Amperes' (this term is normally shortened to Amp or Amps). The unit of Amperes just means the number of electrons physically moving past a certain point in a given amount of time. Lots of electrons per second mean lots of amps, perhaps in the order of many thousands, and not so many electrons per second means a very low density of current, perhaps a few millionths of an amp. These minuscule levels of current are around the same that operate within our own body to send signals to our brain and from our brain to our muscles. Now we know it is voltage and current together that defines the broader term 'electricity'.

We have looked closely at current so I will say a bit more now about Voltage. In an alternative sense, you can imagine this as the force behind the water you see flowing out of a tap when you turn it on. This is a good analogy because you can also think about the rate at which the water comes out of the tap, this being like the current density. This idea works well, because when we open the tap up a little more to increase the pressure behind the flow of water, it gets faster, so we get more water (more current). This is exactly the same as what happens when we consider how electricity basically behaves - varying the voltage up and down will vary the current flow in a similar way. The voltage is the driving force for the electrons, where this can range on Earth from very low values like those inside your body or your mobile phone or computer, to levels that drive lightning strikes that blow trees apart and can send you to the hospital if you are lucky, or to the morgue if you are not!

Voltage levels in space go far beyond those that drive lightning strikes. Here in our biosphere on Earth we have always been safely protected from those much more extreme forces and so are understandably unaware of them. The fact is, the levels of electrical power in space are impossible for us humans to appreciate. This limitation has, however, not stopped the study by some scientists of these extreme levels of energy. We can and do manufacture on small scales in our electrical and plasma science laboratories, very powerful levels of voltage pressure indeed. The person most famous in this area of research is probably Nicola Tesla, a man I will mention again later.

A powerful force indeed! © author's wife

An important thing to point out at this stage is the term 'polarity'. In electric circuits there normally exists a voltage difference between one area and another, a condition that can cause current to flow in a particular direction between those areas. Polarity is what we use to represent the state of voltage (charge difference) that will dictate the direction of current flow. We refer to this charge state by the terms positive and negative, just like when we were talking about charges associated with atoms and their sub-atomic building blocks.

One side of an electric circuit can be more positive or negative than the other side, but this is a relative thing because voltages can constantly vary. If a positive 'node' (point or area) in a circuit varies to become more negative, any other node previously seen as negative in relation to it might then be seen as positive. Try to follow this explanation in the diagram provided here. It is helpful to have an appreciation of this variable relationship in your mind when we talk about current flow, especially if we consider current flow in plasma in space.

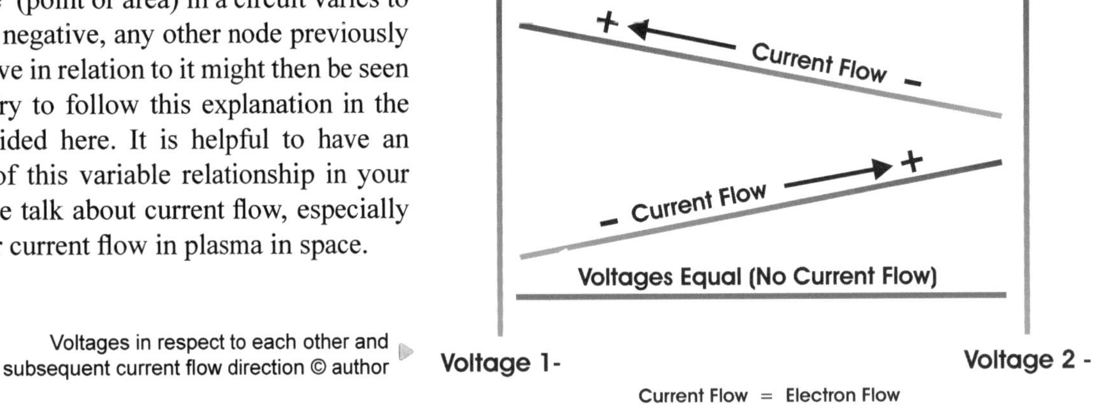


Voltages in respect to each other and subsequent current flow direction © author

For completeness, I must tell you there are two systems by which we refer to current flow; conventional current flow and electron flow. The normal domestic system uses 'conventional flow' which means the flow of electric current from positive to negative. This is opposite to what you will have picked up already so I will explain. The other term 'electron flow' is where we talk about the flow of electrons from negative to positive. The actual effect these terms represent is precisely the same thing, so the only aspect to note is that conventional current flow says that 'holes' (the places where electrons are missing from) flow in the opposite direction to electrons. This can be confusing but it is important when reading about electricity to be aware of this distinction, so please file it away for future reference. For the sake of clarity in this book when I refer to current flow, you can rest assured that I mean electron flow from negative to positive, unless, of course, I deliberately say otherwise.

We now need to add another aspect to our discourse here on electricity, one that relates to the natural effect that electrical conductors have that tends to 'hold back' the flow of electrons through them. Despite what conventional astro-science says, gas in the plasma state in space is not a 'perfect conductor' that has no resistance to current flow. The voltage you could measure at one point of a circuit within plasma is not the same voltage you would measure at another point, so a voltage drop exists that defines a potential difference which therefore allows current flow between the two points in question. When current flow occurs in any conductor (let us stay here with gas in the plasma state) a small amount of energy gets lost through the electrons encountering a 'natural drag effect' within that conductor. This drag or 'resistance' means that energy is being wasted (actually as heat) so it is usually referred to as a 'loss' to the overall amount of energy in the circuit. The resistance within plasma, however, is not very great at all so the losses are low, and so plasma, overall, is considered a very good conductor, but significantly, not the perfect conductor that certain astro-scientists would have us believe.

The exact same thing happens with electric cables due to their resistance in any commercial or domestic setting. Sometimes, the electric energy lost within a cable can be so great due to the cable's internal resistance and the current flow within it being very high, that the cable itself will heat up noticeably, perhaps to the point where it melts its insulating cover or even the copper metal of the conductor inside. This allows us to understand the concept of a thin wire safety fuse in the home situation that vaporises in a flash and bang when too much current passes through it, thereby avoiding damage occurring to some item of domestic electrical equipment.

If electric current can flow in one direction around a circuit, then it can be made to flow in the other direction as well by changing the polarity of the supply voltage. If we did this in the case of a simple circuit with a battery and a bulb, it would not make any difference to the bulb whichever way the current was flowing through it - it would still be at the same brightness determined by the power delivered from the battery. We call this type of steady, single-direction current flow 'Direct Current' (DC) and it does not just apply to batteries, it applies to current-conducting plasma as well. DC usually provides a constant level of electron flow, but this will reduce when the voltage pressure available to push the electrons along gets weak, i.e. when a battery loses its charge. DC power is the form of electricity we will be talking about when we discuss charged plasma in space.

The direction of current flow does not matter © author

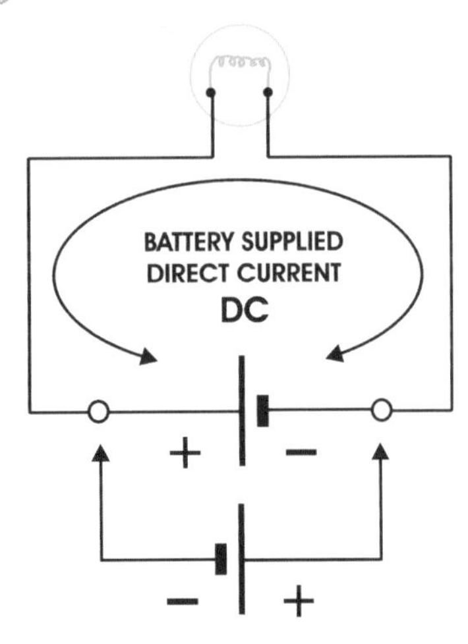


On the domestic scene, we often find batteries acting as sources of direct current and voltage. This is the DC power that operates electric and electronic components within various devices around our home and in most of the portable devices we carry with us. When we later consider how electric power builds up and is stored then released within charged space plasma, you will see that it involves a very similar action to that of a battery.

There is an electronic component worth mentioning that does something very similar to what a battery does; it is also one that works in a way that will assist with our understanding of plasma effects when we discuss them later. This component is called a 'capacitor'. One major difference between batteries and capacitors is that batteries normally supply electric power over a period of time before they run down. Capacitors, however, can 'take in' and 'give out' electric power very rapidly, while storing that power in-between. We will scale this idea up later to a tremendous level when we look at space plasma in greater detail.

There is another approach to current flow that is even more involved in our normal everyday lives. This is where we have 'Alternating Current' (AC). This is the type of current flow that is present in our mains supply at home and all around us in the industry of the modern world.

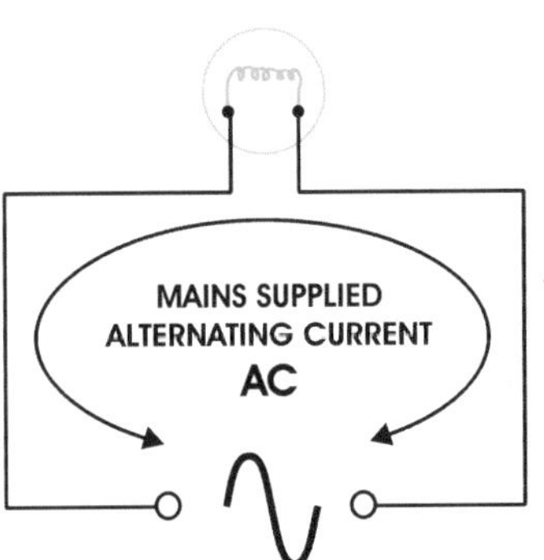


Alternating current goes both ways © author

The direction of AC current flow actually changes all the time by going back and forth along a wire many times in a second. This is because the voltage polarity at the location from where the electric current is being generated is changing back and forth from negative to positive to negative to positive, continuously. This rate of change is 50 complete times (cycles) per second here in the UK and 60 cycles per second in the US. All other countries use one or other of these standards for their mains electricity supply frequency. The idea of having either a steady DC voltage and current or an alternating AC voltage and current is a general one to remember.

Just so that you know the origin of the labels 'volts' and 'amps', 'volts' is attributed to the work of Alessandro Volta, an Italian Physicist, and 'amps' or 'amperes' is attributed to the work of André-Marie Ampère, a French Physicist.

## Magnetism

I sidelined magnetic fields in the previous section to first describe electric fields and the current flow they produce. I now return to magnetic fields to explain how they also bring about current flow.

When we have a conducting path in the form of a complete circuit, such as with either of the two revolving loops of copper wire shown in the diagram here, then we have a route around which electrons can flow communally if some force is applied to make them do that. There will be no flow of electrons in either loop to begin with, but if we introduce the effect of a magnetic field that is moving 'relatively' to those loops, then the free electrons in the copper will be influenced to all move in the same direction around the loops.

Example of how the relative direction of motion of magnetic field or wire conductor gives direction to current flow in a wire © author

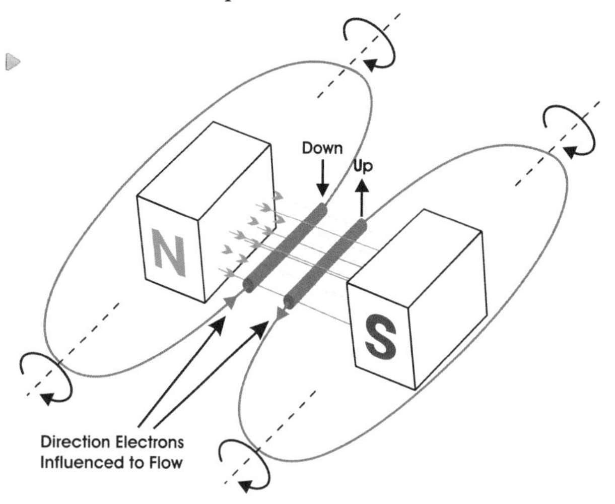


This means that any current flow initiated will have a definite polarity (direction), one that is associated with the direction of movement of the magnetic field responsible. It is important to note here that all directions of motion are linked. If we were to reverse the direction of the magnetic field or the direction of movement of the wire, then the current in the wire would flow in the opposite direction to that which it did before. By changing one, not both, we would have made the current flow first in one direction then the other. This is the concept that was mentioned previously in the description of AC power generation.

The important thing to keep in mind is that to induce current flow (make it happen), magnetic fields and conductors need to move 'in relation' to one another. This is basically how AC power is generated in power stations, but of course, through much more sophisticated and powerful equipment. For our purpose here, which especially involves plasma as the conducting medium, we will be concentrating on situations where, through their dynamic motion, magnetic fields will influence current to flow in one direction only (DC).

If we replace the idea of using a wire with a conducting plasma circuit within a large region of plasma in space, the result would be just the same if we were to come along with a big enough magnetic field. Magnetic fields in space are dynamic (ever-changing) and they are to be found all over the place (they are ubiquitous). They are especially concentrated around bodies such as galaxies, stars, planets and in and around filaments and concentrations of plasma. Any images this may bring to your mind, especially on a grand scale, will be helpful for what is to come, so hold on to those images.

We experience two types of magnetism, electromagnetism and the plain old type in a normal permanent magnet. Electromagnetism, as a concentrated form of the EM force, can be generated by the flow of electric current through a conductor. It presents itself as an invisible field around any conductor that has current flowing through it. Depending on the direction of this current flow, the magnetic field will also have a direction to it, one that is directly associated with the direction of the current responsible.

To understand this relationship, there is a simple rule that electrical engineers learn, known as the right hand grip rule. If you hold your right hand with your thumb pointing in the direction of *conventional current flow* in a conductor (that is, positive to negative) then the natural curl (direction) of your fingers will indicate the direction of the magnetic field that is formed around that conductor.

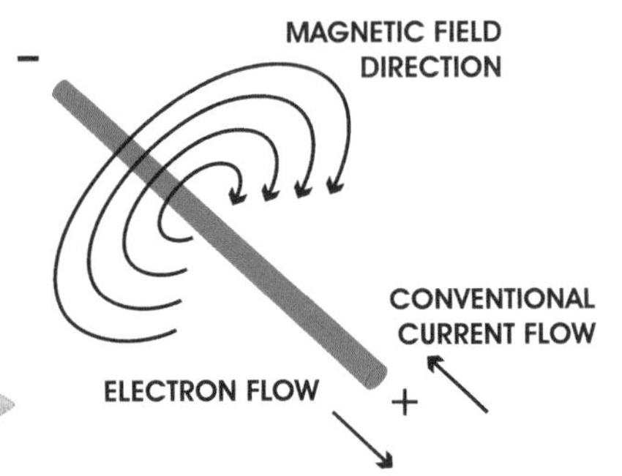


Current direction and magnetic field direction relationship © author

Normally the strength of the magnetic field around a single wire conductor is not very great, so to make it more powerful and useful, we wind a long length of wire into a coil so that a concentrated, more powerful magnetic field forms through the addition of the fields produced individually by each turn of wire. This is how we make powerful magnetic fields that operate inside electric motors, in starter coils in motor cars, in television tubes and to keep Maglev trains floating in the air. Note here that we do not see EM fields in our everyday modern lives but we certainly can see and experience the results of their effects.

Just like electricity, magnetism has polarity associated with it; that is, electricity has positive and negative polarities, and magnetism has its equivalent 'north' and 'south' magnetic polarities. A magnetic field's flow (or direction of influence) is considered to be from north to south, or 'N to S' as we often find in text descriptions. You will recall that we said electric current flow is where free electrons are attracted to atoms that have positive 'holes' available due to one or more of their electrons being missing. Well, in the case of magnetic fields, polarity also determines attraction and repulsion. This is where we have another simple rule to help us remember what is going on; magnetic poles that are not the same will attract but poles that are the same will repel. In other words, N and S and S and N will attract, while N and N and S and S will repel.

**Electro-magnets** are able to concentrate the strength and direction of magnetic fields. One example of this is in a car, where we find an electro-magnetic component that plays an important role when the key is turned to start the engine. When we do this, current flows through a copper wire coil inside that electro-magnetic component (usually referred to as a 'solenoid'). This current forms a strong magnetic field within a metal core, around which the solenoid's wire coil is wrapped. The electro-magnetic field produced in the metal core then attracts towards it a pivoted metal plate that has a heavy duty electric switch mechanically linked to it.

Car starter switching circuit © author

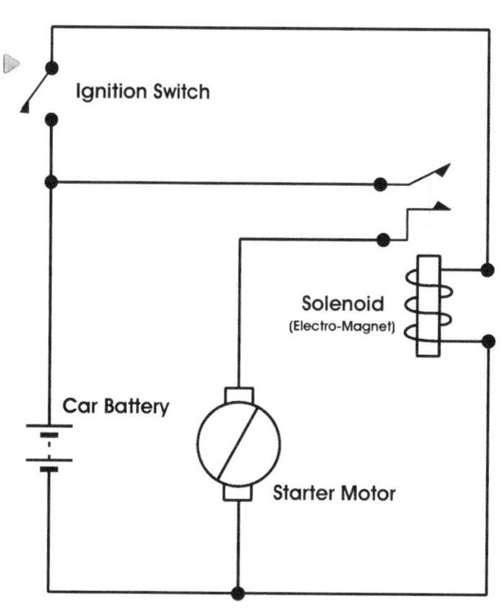


When operated, this switch allows a heavy current from the car's battery to flow to the starter motor which turns over the mechanical working parts of the engine at the beginning of the engine's start-up process. One important thing to be aware of here is that the metal core that became magnetic (magnetised) by the coil, was made to assume a particular N-S polarity (north at one end and south at the other) and that this effect was due to the DC from the battery flowing through the coil in one direction only. This means we can consider passing current in either direction through an electro-magnet's coil to make it produce both N-S or S-N polarities, as required. Depending on the mechanical (or electrical) effect we want to achieve, this can even produce the force of both magnetic attraction and repulsion in relation to some external piece of equipment. There are clever ways in which we have discovered how to employ this effect, certain types of electric motors being one, and bells, electric locks and electro-magnetic brakes, being others.

The other type of magnetism we need to consider is the 'fixed magnetism' of a 'permanent magnet', the kind that does not involve the flow of electric current and which is commonly associated with the ornamental paraphernalia we stick on our fridge doors. Some materials, usually metals or composite materials with a metallic content, are easily turned into permanent magnets by subjecting pre-shaped forms of that material to an external and very strong magnetic field for a period of time. This process influences the atoms within the material to take on a communally oriented direction that remains in place, giving the material an overall magnetic polarity, until such time as the communal orientation of the atoms breaks down and the magnet loses its strength.

This ability of metals and compound materials to retain for some time the 'permanent' magnetisation they have been endowed with, varies in terms of how magnetised they can be made to be, and for how long that magnetisation can be retained. In addition to this being a description of the basic manufacturing approach to creating permanent magnets, we now become aware that some materials are better at retaining their magnetisation than others. For instance, steel is not very good at retaining magnetism but iron is much better, and the material known as neodymium is exceptionally good at retaining its magnetisation. Neodymium is known as a 'rare-earth' material that is used for many of the exceptionally strong magnets we see on the market these days. There are some naming terms associated with the magnetic properties of material. 'Permeability' is the ease by which materials are magnetised, 'remanence' is the ability to retain the magnetised state, and 'coercivity' is the difficulty with which de-magnetisation is achieved. We do not need to remember these for our purpose here.

Getting back to electric currents and magnetic fields, we can see that a fundamental association exists. This is where electric currents can produce magnetic fields and magnetic fields can produce electric currents. These two things are inextricably linked; in fact, one extremely important aspect of this should be noted and remembered: this is that magnetic fields can only be formed by the flow of current within a conductor, or in other words, electric current flow is essential for magnetic fields to exist.

We can now take another step forward by considering more about the movement of a conductor through a magnetic field, or in opposite relative terms, a magnetic field passing 'through' (across) a conductor. In a DC generator, any current made to flow will be in one direction only, and that current will exist for the same time that the relative movement between the conductor and the magnetic field is maintained. This is how electric generators work, where coils of copper wire are spun at high speed within a magnetic field (the spinning coil possibly being driven by the flow of water, wind or steam in a turbine, or by a diesel engine).

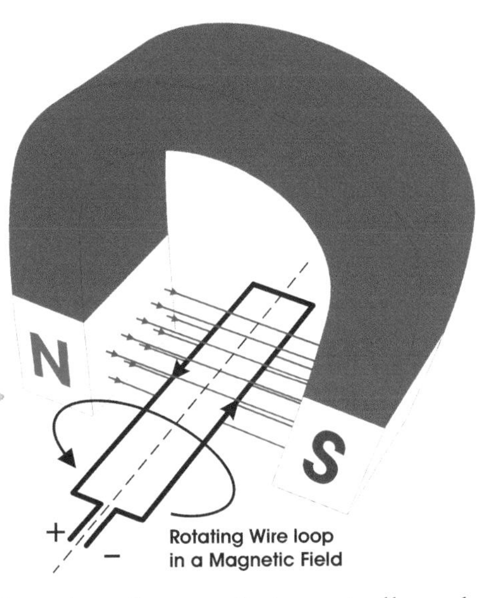

Basic DC Generator © author

It is through a very similar process in our national power stations that AC power is generated by the rotation of a particular configuration of copper coils in magnetic fields. The resulting alternating voltage produces the alternating current that flows to the outside world by way of the grid distribution network and a series of high to low voltage transformers that eventually ends up with the AC mains power we use in our homes. This is the fundamental process by which we receive the domestic electricity supply upon which we depend.

Magnetic fields are intrinsic to our explanation of an Electric Universe, where currents flow through regions and filaments of space plasma due to the potential differences that exist within and between them. When those currents are on the scale of the gigantic plasma filaments we are now able to observe, magnetic fields of exceptional power are formed which surround and further constrict those filaments into the often twisted forms we see. Due to the dynamic nature of plasma regions and filaments in space, the never-ending interaction of their magnetic fields induces further current flow in regions and filaments of plasma present within their vast neighbourhood. This is the basis upon which I will develop my description of how electric power is constantly being generated and brought into play within the environment of space.

**The four fundamental forces**

Having previously mentioned the four fundamental forms of matter, I will now describe the so-called **four fundamental forces** that are commonly discussed. These together are said to account for how we perceive and interact with our reality and how our Earth and space environments appear to us as they do. As I cover the following information, I would ask you to try to think in both really big and really small terms.

*A 'force' is any influence that attempts to change the motion or shape of something.*

The four fundamental forces that we are told exist are;

- The strong nuclear force
- The weak nuclear force
- The electromagnetic force
- The gravitational force

**The strong nuclear force**. This binds protons to protons and neutrons to neutrons and protons to neutrons, to form the central nuclei of atoms. Protons would not normally want to be together because they both have the same charge (remember, like charges repel). Therefore, the strong nuclear force at that scale is what somehow overcomes that reluctance to be together and keeps all these particles bound in the form of an atom's nucleus. It is also said to be the force that binds the internal structure of protons and neutrons to give these sub-atomic particles their own form. So, in this incredibly small-scale setting, the strong nuclear force is seen as an extremely short-range but exceptionally powerful force. It is the force responsible for the energy involved in atomic explosions!

**The weak nuclear force**. This, as its name infers, is weaker than the strong nuclear force, but the job it does is also at the sub-atomic level. The weak force is responsible for a naturally occurring event that takes place when, for example, a neutron in the nucleus of an atom 'changes itself' (decays or reverts) into a single positive proton and a single negative electron. When this happens, some amount of energy that is no longer required is also given off as radiation. In other words, the weak nuclear force is responsible for certain elements decaying from one element into another through either the conversion of neutrons into additional protons inside the nuclei of their atoms, or the release from atoms of groups of particles consisting of protons and neutrons. An example of decay would be if you had some uranium and left it for a very, very long time (hundreds of millions of years). You would eventually find that much of that uranium had turned into lead. In this sense, uranium is known as an 'unstable element'. This label is applied because certain elements release energy over time as radiation through this decay process to eventually become a different, and often 'more stable' element. Natural decay of elements is a quite common occurrence and it is generally not a dangerous thing for us humans, but it sometimes can be - it all depends on the actual element involved and the amount of energy released by its decay process.

Here, the often powerful radiation generated by the decay process can damage the cells in the tissue of our bodies, just as powerful radiation from any other source can similarly bring about. You will no doubt have heard of people getting radiation sickness or that radiation has caused skin burns and cancers: well, this is essentially how these types of tissue damage can happen. Note here in terms of the decay process that this is an effect that arises from the action of the weak nuclear force and not a direct effect of that force itself. We humans do not experience either the strong or the weak nuclear forces in our daily lives due to their range of effect being only on the scale of an atom's nucleus. Make no mistake though, these forces involve levels of power that are well beyond our normal ability to imagine!

**The electromagnetic (EM) force**. This force gives us the structure of the atom itself by binding electrons in their orbital shells around the central nucleus and it also, perhaps surprisingly for some, allows us humans to experience and interact with what appears to be the 'solid environment' in which we live and operate. It is the force at the heart of 'our impression' of the sensation of touching a solid surface and other forms of tactile contact with the physical world. It prevents our hand from passing through a sheet of metal and it stops us from falling through a wall if we lean against it. It is the force at the root of what we see with our eyes and consider to be real and so it helps us make sense of the world we live in, and it is the force behind the gamma rays used in medicine to kill cancer cells in our bodies. These examples, and a myriad more, have fundamentally the same force of EM at their root. It is a two-way force; it is attractive and repulsive and it is also a long-distance force, its influence being over an infinite distance. It is the force that rules chemistry and chemical reactions through the bonds made and broken between atoms and molecules. In terms of our current understanding, it is the EM and gravity forces with which we humans are in direct regular contact every day of our lives.

**The gravitational force**. Gravity is described to us as the force that keeps all matter on large scales together to different degrees, depending on the amount of mass the matter involved actually has. Its influence is long-distance and extends over an infinite range, just like EM. It is said to be the force responsible for everything of a physical nature we can detect in space and for keeping our universe and everything in it together in the forms we observe. In the human experience, gravity is a one-way force of attraction; it has no known repulsive property. If you take a moment to consider how you 'feel the weight' of your body being pulled down as you sit or stand or walk or run, then this is the effect of gravity that you are experiencing. The original law of gravity was presented to the world by Sir Isaac Newton, and it is interesting to note that his law only informs us of the effects of gravity and not how gravity itself actually works. This lack of a total explanation is interesting and relevant and should be kept in mind for the information to come. For now, I am comfortable with my personal belief that gravity is also a manifestation of the EM force, just as the strong and weak nuclear forces are as well; we just do not yet fully understand this all-encompassing role of the EM force.

It is all very well for some to sound confident when they talk about the forces we find in nature, but the fact is, those forces and their relationships remain enigmas to science. We can talk about the effects of any force but we still cannot describe in clear terms what, most fundamentally, a force actually is. Despite this, the ultimate goal of science is to do away with these four separate definitions and to have only one force defined and accepted as being at the root of everything. For me, this will be the force of electromagnetism, but this view will not be shared by many in today's gravity-centric astro-science. Could a clue to this be that astro-science has not yet looked in detail with an open mind at what electromagnetism is about or the part it may play in our universe?

We need to remember that the important things we have been told about gravity have not been proved and that what we have been given is at best, only opinion. The fact is that better theories are available to explain the form and operation of things which the force of gravity has traditionally been taken to explain. These alternative theories have been around for many years but have been largely ignored, sometimes suppressed and frequently ridiculed by those who either do not understand them or who have other agendas. These reactions are professionally immature, inappropriate and come on occasion from those who seem to have a stake in protecting either their own careers, their projects, their egos or their status within some 'valuable' hierarchy.

So, to reinforce the relevance of plasma in space and its link with the EM force, and to 'lift the curtain' further on what is shortly to come, I need to mention how the effects of plasma present themselves in the environment of space. 'Space' is aptly named because there sure is a lot of it! It is bigger than we can imagine and it is filled throughout with mostly dark mode plasma that is concentrated in regions of various dimensions, constituent elements and densities, all of which have positively and negatively charged regions within and around them.

It is truly hard to imagine the immense power that these vast regions have stored within them, untold numbers of which are millions of light years in size. Over time, these mostly invisible but now detectable super-galactic-sized clouds and more obvious formations of charged matter interact with each other and with regions of neutral matter (dust molecules and other gases) as well. This takes place through electrical and magnetic events that induce current flow within and throughout all their various forms. In turn, the gigantic currents produced generate further magnetic fields, which being dynamic in nature (i.e. continually moving and interacting), go on to induce additional current flow in adjoining regions, sheets and filaments of plasma.

This cycle of interplay goes on and on, and as far as we know, plasma in space has forever been there going through this same process. When powerful currents flow within plasma, the most common appearance they take is that of long and twisted filamentary strands ... Why does this happen? Well, we know that current flowing through any conductor produces a magnetic field around it in a circular formation like a tube. This is just the same with plasma when it is conducting current, and this is the reason we see filaments of plasma, because they are being constricted by their own magnetic fields into that strand-like form. Another action of the naturally attractive magnetic force is to bring these filaments together in pairs to form a helical twisted appearance that looks as if those pairs of conductors are 'entwined' with each other.

This twisted form will come as no surprise to some because it is the natural and most electrically efficient way to retain and transmit energy across great distances - we know this from basic electrical engineering. On the scales in question, these twisted pairs therefore act just like the power lines of our national grid distribution networks. They instead form the cosmic energy distribution network that supplies power to galaxies, stars and other structures on the grandest of scales. This idea is key to picturing the process by which galaxies and stars are formed and how they go on to behave in the various ways we now see they do. We shall be looking much more closely in chapter six at how these things are actually achieved.

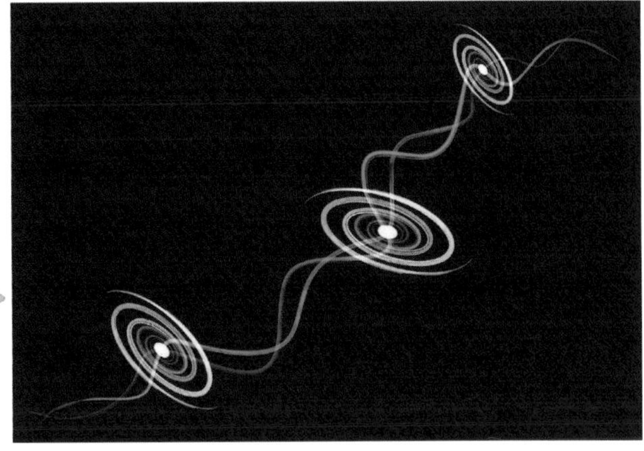

The idea of galaxies strung along a galactic power line © author

Go back to the idea of our inability as humans to really appreciate the scale of things in the universe and keep in mind that it consists of plasma to the tune of 99.99%.

From our everyday experience within this thin biosphere of breathable air, we modern humans have previously never had reason to focus on the evidence around us that indicates powerfully charged plasma exists in space. The reason for this is the highly unusual physical separation that exists between our living environment on the surface of Earth and the vast environment of space. This implies that our main experiences only relate to the other three types of matter when they are in their neutral state; plasma being obvious to us only in the forms we produce for our own purposes such as fluorescent lighting, electric arc welding or where we see it in nature as lightning and fire. These forms of plasma do not constantly or in any overly intimidating way, surround and impress us like the powerful manifestations we experience of wind, water or solid ground eruptions. This is why the powerful but small-scale effects of current-carrying plasma in our everyday environment, have been relegated and only considered useful at our own scale of experience through man-made processes. Beyond Earth's biosphere, the actual situation could not be more at odds because plasma is the most fundamental form of matter and it exists at every scale from sub-atomic to cosmic. This is where our notion of scale can help us to appreciate what is really meant by the implications of 99.99% of our universe being made up of plasma.

Our universe is so big that it is impossible for us to hold an adequate appreciation of its size in our minds. At our present level of mental capability, how can we hope to develop a true feel for how big and old things are - how small they can be - how fast things can be - how far away things can be and especially, how powerful and effective electrical energy in space can be. The universe is unimaginably big but we must also think about the sub-atomic structures of all forms of matter which are unimaginably small.

The relative location within this scale of things where we humans experience reality, is a very tiny place indeed; it is somewhere in the middle between the two extremes. Attempting to study these vast differences in scale has been very enlightening for me and it has helped bring a measure of clarity to my perspective on everything.

It seems (to me) that to claim what astro-science has already seen fit to claim about the crazy enormity and operation of the universe from the position of our couple of hundred years of modern experience and learning within our tiny biosphere, is just silly and typically arrogant of us. All I believe we really can do is to build worthy research programs from science that we already know works, then from what we discover, form further impressions of our reality with which we can live until we refine these through similar good science.

The 0.01% not included in the 99.99% figure for plasma in the universe is a token amount that represents the electrically neutral matter left over, such as what we have in Earth's biosphere, after taking into account all the other forms in our universe that are in a charge differentiated state. Remember here that the EM force is overwhelmingly more powerful than the force of gravity. So, since we are talking about charge differentiated plasma on galactic and stellar scales that carries enormous electrical power, we are reminded yet again to question the 'do-it-all' claims that the puny force of gravity has been associated with.

### The EM Spectrum

This is one of my favourite areas for it touches so many aspects of our lives and acts as a foundation on which to build an understanding of a great many important things in our environment. Here I have in mind that we go through life not usually paying too much attention to the objects that surround us and the natural events that occur; things such as …

- What we physically feel when we enjoy a bright sunny day.
- How music is received and played to us by the radio in our car.
- What takes place when we go to hospital to get that sprained wrist X-rayed.
- What burns our skin if we stay out in the sun for too long.
- How we produce images from space that would otherwise be invisible to our normal telescopes.
- How our microwave oven bakes those potatoes.
- Why our toast burns if we leave it under the grill for too long.
- Why everything, especially our natural world, appears so colourful to us.
- How the military and security folks have the ability to see in the dark.
- And, how we can control tiny robots on the surface of Mars from here on the Earth.

These are all examples that involve the influence of the energy represented by the EM spectrum. Its range is enormous and the EM radiation it represents influences and is responsible for so many things and events in our everyday lives; many of which we take for granted and therefore have no idea of what lies behind them.

Not too long ago I saw the EM spectrum described in a very thought-provoking way in a television science documentary. An estimate for the length of the total Electro-Magnetic Spectrum (EMS) was represented by the distance between the Earth and the Sun which is 93 million miles. It turned out that the length within this EMS representation that would be the section of visible light our eyes can detect, would be only 8 inches long. Think about it, 8 inches out of 93 million miles. Wow! In terms of the full EMS this is a very insignificant fraction that represents absolutely everything that our eyes are limited to seeing! What if our eyes could see more of the EMS than they do, what would it be like if our vision went beyond that tiny 'visible light' portion? Thinking about this lets us appreciate that there exists all around us much more invisible EM information, which if we could see it, would reveal a very different picture of our world and our surroundings in space than we are used to. Remember the old film … "The Man with the X-ray Eyes"? Well, science and technology has provided the means to give us what that science-fiction adventure suggested, so we are now able to see (observe) with our equipment, much more than our eyes alone are capable of, not just on Earth but out in the universe too.

The EMS stretches from the lowest form of radio signal right up to the most energetic and deadly gamma rays that mostly come to us from deep space. Although high-energy radiation is harmful for us, we generate it here on Earth with equipment we build for medical, scientific and commercial uses. In our everyday lives we are constantly exposed to some form of EM radiation to some degree or other. Most of the frequencies (wavelengths) of radiation with their typical power levels, are harmless to us here on the surface of the Earth. However some, like ultra-violet light, X-rays and gamma rays, will in certain circumstances destroy human tissue.

The EMS has a fundamental link with frequencies at the sub-atomic level for all elements and with the energy levels these frequencies represent. The thing to keep in mind here is this fundamental link between frequency and energy, where the way this works in everyday life is that we have low frequencies with low energy and high frequencies with high energy. As an example from both these extremes, we have sensitive radio receivers capable of picking up weak low frequency radio signals at the bottom end of the spectrum, then we have gamma rays at the high end that would smash the atoms in our bodies apart if they were to strike them.

Another term, which for the sake of being complete I will now say more about, is 'wavelength'. This is directly associated with 'frequency' where one complete wave (one wavelength) is the same as one complete cycle of frequency, or to bring in another term, one full 'oscillation'. The most common form of a wave is a sinusoidal wave or 'sine wave'. Sine waves normally occur in a constant series of complete cycles or wavelengths, the distance between any two similar points being termed the wavelength of the cycle.

A wavelength is measured in metres, centimetres, millimetres, micrometres, nanometres, picometres, femtometres or attometres and its associated frequency is normally measured in Hertz (Hz). For instance, where we talk about having a radio frequency of 145Mhz (which means 145 million cycles, wavelengths or oscillations per second) then this frequency is said to have a wavelength of around 2 metres in length. This is saying that because the radio wave is travelling at the speed of light, 300,000 kilometres per second, it will advance a distance of approximately 2 meters as it completes one full cycle of frequency (one wavelength or one oscillation). Here we have introduced another important thing to remember. The EMS has a direct relationship with the speed of light, where all EM radiation travels at 300,000 kilometres per second - in a vacuum.

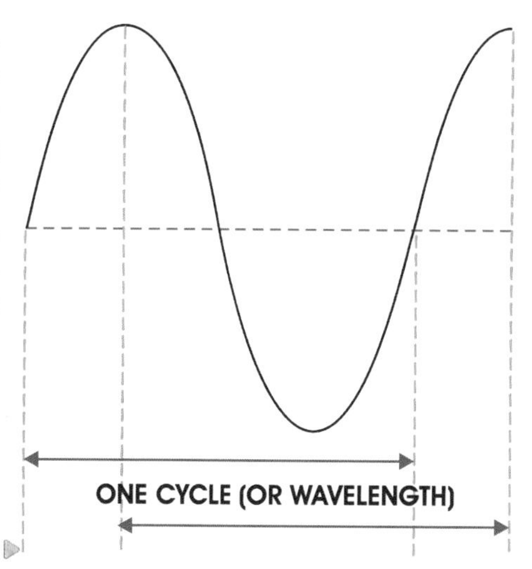


A Sine Wave and the "Cycle" and "Wavelength" relationship © author

The visible spectrum, the tiny spread of frequency that goes from red through all the other colours to violet, is all from the EMS that we humans are visually aware of. The colourful world we are used to seeing is therefore a limited view of what is really available for we fool ourselves into believing that 'what we see is all there is' - a notion that, as I have already said, is so very far from the truth.

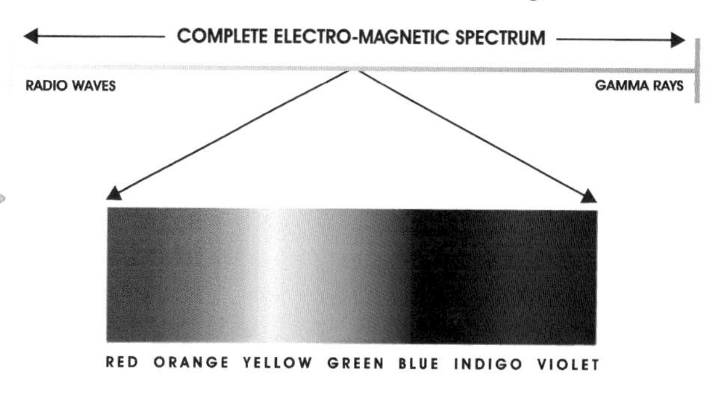


The visible portion of the Electro-Magnetic Spectrum © author

All EM radiation is photon energy, the production of which we know about from our previous discussion. Our senses other than sight and touch do not have a direct relationship with the EMS; for instance; smell and taste are chemically based and hearing is mechanically based around variations in air pressure. Touch might be thought of as mechanical but it can also be considered as more. Here we have the sensation of temperature awareness which is actually the sensitivity of the nerves in our dermal layers to infra-red (heat) radiation. This should put our senses in some relationship with the EMS.

We are happy with the senses we have because the information they provide us with is all that we have ever been aware of. On this basis is easy to see that we are not that well equipped by our five rather limited senses to detect the wider presence of EM radiation in our environment. Here are some examples of EM radiation that exist around us or of which we make use on a daily basis ...

- Radio waves for communications
- Microwaves for cooking and communications
- Radar for safety, security and weather
- Infra-red radiated heat for all forms of warmth and comfort
- Illumination of various sorts
- Ultra-Violet light for revealing things normally invisible and for getting a tan from the Sun
- X-rays that show cracks in metal fabrications and allow pictures of our bones to be produced
- Gamma rays that kill cancer cells and destroy harmful bacteria

So there we have it: absolutely everything that can be called EM radiation comes about through electrons being forced to temporarily absorb energy then release it in the energy packets we call photons. These released photons then have particular frequency and energy attributes that are associated with the energy level and type of matter they originated from.

### Nuclear Fission

This is the process by which an atom's nucleus is made to break apart and release great amounts of energy in the process. Normally, this involves a certain variety of the element uranium which when used in today's nuclear fission reactors is responsible for generating electrical power in a large number of countries around the world who have chosen to adopt this process.

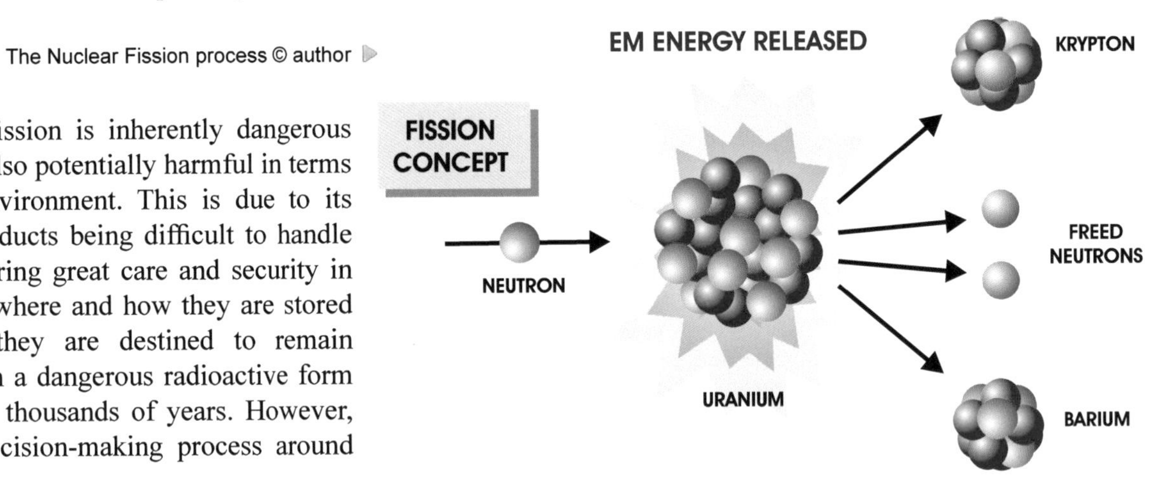


The Nuclear Fission process © author

Nuclear fission is inherently dangerous and it is also potentially harmful in terms of the environment. This is due to its waste products being difficult to handle and requiring great care and security in terms of where and how they are stored because they are destined to remain with us in a dangerous radioactive form for many thousands of years. However, as the decision-making process around fission power reactors is centred around finance, they are popular because they are cheap to run. This is fine when they are operated properly, but the fact remains that the fission process is definitely risky and a very costly one in more ways than just financial.

In fission reactors the process involves a type of uranium, the atoms of which are struck by high-energy (high-speed) neutrons that split the nuclei of those uranium atoms apart. When this happens the sub-atomic particles from the original nucleus fly apart and there is a release of energy that is no longer required to hold things together. Further neutrons are released during this process that go on to collide with yet other atoms of uranium; and so the same process is repeated in a very rapid cascade-type fashion. You might imagine this chain-reaction could run away with itself and go out of control, and you would be correct, so it has to be deliberately regulated. This is done by employing carbon rods to absorb some of the powerful sub-atomic particles that are flying around. When these carbon rods are deeply inserted inside the reactor core, the overall energy generated will be at a relatively moderate level. As the rods are then withdrawn, more interaction is allowed between neutrons and the atoms of uranium, so a greater amount of energy is generated. The useful energy released in the form of heat is then used to boil water that produces steam which in turn drives turbines connected to electrical generators that eventually produce electrical power. This is what goes on in our nuclear power stations today.

## Nuclear Fusion

You can look upon nuclear fusion as being (at the atomic level) the opposite of fission as it is the process where atomic nuclei are forced together rather than split apart. Nuclear fusion has been presented to us as the process that allows our Sun to work by changing a form of hydrogen gas into helium gas and releasing great amounts of energy as it does that.

The Nuclear Fusion process © author

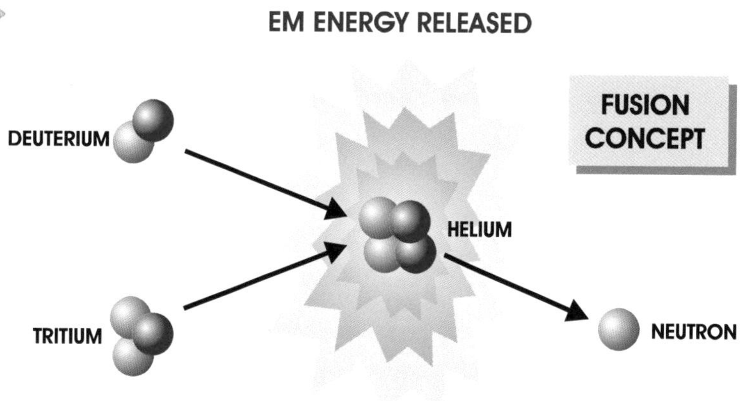


The current thermonuclear model of the Sun says that it is a gigantic ball of hydrogen gas and that nuclear fusion is taking place at its core. Here, the fusion process is due to forms of hydrogen nuclei being compressed by extreme physical pressure and agitated with tremendous temperatures so that they fuse together to form new nuclei of helium. During this process excess neutrons are released. To further explain this ...

Helium is a gas with 2 protons and 2 neutrons in its nucleus and 2 orbiting electrons, whereas hydrogen's isotopes of deuterium and tritium have a single neutron difference between their nuclei, and a single electron each - this leaves one excess neutron when an atom of helium is produced. This process of changing one element into another releases extremely high levels of EM radiation such as X-rays and heat, because once again, not all the energy involved in the process of creating the end result (helium) is thereafter required so the photons of left over energy have to go somewhere. In terms of our own star the Sun, it is presumed that after many hundreds of thousands of years of bouncing back and forward within a peculiar 'radiant zone' inside the Sun, these photons lose energy and eventually make their way to its photosphere 'surface'. This photon energy is then radiated away into space as the heat and light we experience here on Earth. If this is correct, then it would mean that the light and heat we find coming from the Sun today was actually generated hundreds of thousands of years ago at its core. This, essentially, is what we are told is going on inside our own star. We will be returning to the idea of nuclear fusion-powered stars later to delve into why this, in fact, is not the way the Sun works and to take a close look at how all stars actually do work, electrically of course.

**Distances in space**

You will probably have heard somewhere or other the term 'Light Year' being bandied about. This relates to the distance that light will travel in one year at the speed of 300,000 kilometres per second in a vacuum. This equates to 5,878,499,810,000 or almost six trillion miles in that one Earth year. If we think about the age we have been given for the universe, 13.7 billion years, you can see that for each of these years, light will have travelled a distance of approximately 6 trillion miles. If you wanted to, you could roughly calculate the diameter of the universe if you multiply these two figures together and double the result - 13.7 billion x 6 trillion x 2. The end figure is a very large number of miles indeed, but this wouldn't be a useful figure because during the 13.7 billion years that have already passed, light will have travelled the same distance again and an acceleration of this expansion will also have been involved - well, according to standard theory, that is!

Closer to home, the star nearest to us known as Proxima Centauri is 4.3 light years (ly) away. This turns out to be a little more than 25 trillion miles. Here I have summarised an understandable model which illustrates the distance our own Sun is away from Proxima Centauri. I have taken this from Don Scott's book "The Electric Sky" but it was originally put together by Robert Burnham Jr., an American astronomer ... "*Consider the distance of one mile. At one end we have our Sun the size of the full stop at the end of this sentence and the Earth one inch away from it as an almost invisible speck of dust. Pluto's distance from the Sun would be around three and a half feet away and Proxima Centauri would be another full stop, four and a half miles away!*" There is a further interesting point to consider here. The gravity model of the universe tells us that everything is subject to gravitational relationships. What then do you think the gravitational relationship will be between two specks of burning hydrogen the size of full stops that are four and a half miles distant from each other?

Much further away than Proxima Centauri we have our nearest major galaxy Andromeda at 2 million ly. That is around 12,000,000,000,000,000,000 or 12 billion, billion miles away; not by anyone's standards a Sunday stroll after lunch! Everyday use of enormous numbers like this is not very practical and this is why astronomers frequently use the unit for light years (ly) and another one (AU) that represents 'Astronomical Units' ... So then what is an AU? One AU is the distance between the Earth and the Sun; 93 million miles. The AU is a much shorter distance than the ly but their use is for the same purpose of representing unwieldy numbers in a more convenient way. The term AU is normally only applied on the smaller scale of the Solar System itself. For instance, the mean distance from the Sun to Mercury is around one third of an AU or 0.39 AU or 36 million miles, and the mean distance to Pluto is 39.53 AU or 3,647 million miles or more than 39 times the distance that Earth is from the Sun.

The information we have covered at a basic level here in chapter four is relevant to the rest of the book. You are close now to being well enough armed to read and consider what is coming and hopefully then apply your new understanding to coming to terms with why things in our universe really do work in the way the Electric Universe model suggests they do. First, however, we must consider some additional relevant characters and their important work in order to complete our database.

# 5 | The work of the honourable but ignored

Any introduction to the Electric Universe model could not adequately be attempted without mentioning the people whose work has contributed to the theories now fundamental to that model. Some of them are big names in the history of science, others are less well known and some continue to expand and refine that model through their contributions today. The reaction by many in mainstream astro-science to this body of people and their work has been unfortunate, where some have been ignored and ridiculed and a few even intellectually attacked on occasion. This chapter documents the main characters whose work I have come across and the contributions they have made to the evolution of the EU model. It also puts the final elements of our build-up in place before we move on to look at the basic theories of the EU model.

As people you may already be aware of, I will first mention Galileo Galilei, Isaac Newton, William Herschel, Johannes Kepler, Albert Einstein, Edwin Hubble, Fred Hoyle and Steven Hawking. These are just a few of those whose work in astro-science and mathematics has contributed to the theories that lie behind the current standard model. I have respect for every one of them and believe they did good honest work by contributing as best they could to the advancement of science. However, I have taken particular note that in the opinion of many experts, some of the hard work and creative ideas of these people have either turned out to be wrong turns in themselves or have been distorted through work carried out by others in their own pursuit of achievement.

I now mention the names of James Clerk Maxwell, Nicola Tesla, Kristian Birkeland, Irving Langmuir, Hannes Alfven, Immanuel Velikovsky, Charles Bruce, Ralph Juergens, Earl Milton, Halton Arp, Anthony Peratt, David Talbott, Wallace Thornhill and Donald Scott. I would not be surprised to find that most of these names are unfamiliar. This chapter expands, to a degree, on the significant work that some of these people have done and how outcomes from that work now support the EU model. For now, I will say that with similar honest endeavour, dedication and creative thinking, these people have contributed significantly to research and discovery within electrical and magnetic science, plasma physics, cosmology and astronomy. You will note that mathematics is not included here. This is because it is viewed by supporters of the EU model as mainly a subordinate proving tool to observation and experimentation and to the conclusions drawn from those activities. This stands in contrast to the way today's astro-science has come to view the subject of theoretical mathematics. Instead of using it as a proving tool for obtained results, they have for many decades given it a leading role, an attitude that has encouraged an environment where the formulation and acceptance of impossible theories is commonplace.

The status that mathematics has in the astro-science world is an evolved one that has in many ways turned out to be truly damaging. It seems that too much belief has been inappropriately placed in theoretical mathematics as a first-stop-shop for ideas and theories. As a result, not only have serious fundamental questions arisen about the direction and conclusions reached by today's astro-science, the big picture seems also to have become overly complicated and opaque to common sense. Things are further complicated because in their attempts at proving the relevance of this confusing situation, even more of their imaginative mathematical constructs have been added to it. This has resulted in a confusion of incompatible assumptions that have taken up the thinking time of good scientists - all to no end other than to cause division and build walls between these people and their work. Many of the standard model theories that have emerged from this situation have been shown to be impossible to work in the real world or to test in laboratories using proven methods - yet still they have been accepted and linger on, just like a bad smell, and astro-science has not been called to account for this state of affairs. In short, mathematics is regarded differently between the mainstream astro-science and EU communities. One side puts theoretical mathematics first and builds its theories from it, and the other side places it secondary to logic, common sense and actual results of observation and experimentation. Here, it would not be a big challenge to guess correctly the nail on which I suggest, a fair minded person's coat should hang!

The division that this difference in approach highlights is further added to by the current lack of cooperation within and around science disciplines in general. There is no doubt that scientists working on their own have the chance to achieve good things, but the quality of their work and their rate of output would improve if they were open and confident enough to discuss what they are doing with colleagues who could help. Although a logical thought to have, this turns out to be naïve, for an open attitude to work is rarely practiced by today's science research communities. Instead, we have significant isolationism between disciplines, usually for commercial patent protection, egoism and project funding (survival) reasons. Ideally, of course, this should not be the case. If there is even a chance for good cross-discipline cooperation to take science forward, then this is what should be done. In recent years, many 'maverick scientists' have deliberately kept themselves away from being involved with the mainstream attitudes I have described here. These are mostly scientists who talk and cooperate across their disciplines and who are therefore seen as part of what has been rediscovered from earlier times as the 'interdisciplinary approach to science'. Hopefully, this is the way research will again be done in the future.

It is interesting to note that in the very early days of research, unlike today, scientists were able to choose to work on whatever took their fancy. This was a situation without any undue external pressures from the early science institutions, universities and whatever commercial interests existed at the time. There was a genuine sense of proper intellectual decorum around this, because these people truly believed that what they were doing was for the overall betterment of science and mankind and that the sharing of information would help work towards that goal. There are examples of early scientists making extraordinary personal efforts and sacrifices to prove their thoughts and ideas correct, according to their own interpretation of the rules of rigorous scientific research. The dedication displayed by many was often admirable, and in some areas at least, it has remained this way with the more honourable types. We now look at some of the individuals from history and from today to whom we should be beholden, in my opinion, in terms of the advancement of general science and for the original contributions they made and which some still make to what has become the Electric Universe model.

**Benjamin Franklin (1706 – 1790)** was the multi-talented American politician, musician, philosopher, scientist and inventor who was one of the first to think about the broad presence and influence in our world of electrical power. He had a particular interest in lightning and believed that it was an electric force. His famous kite flying experiment which took aloft a metal key during a thunderstorm to see what would happen, was a practical attempt at proving the link between lightning and electricity. This however was in the early days of electricity, and many years would pass before the Norwegian scientist Kristian Birkeland really started the ball rolling on the search for electricity in the atmosphere and in space and for the effects that phenomenon has on the Earth itself. It would be from Birkeland's research that the discipline now referred to as 'Plasma Science' would emerge.

Benjamin Franklin

## Professor Kristian Birkeland (1867 – 1917) [5-1]

This man trekked across the dangerous frozen arctic wastelands to perform experiments that involved detailed studies of the aurora, or Northern Lights as we call them today. His theory was that these wonderful dancing displays of atmospheric light at Earth's high latitudes were fundamentally caused by electrical currents coming from the Sun which interacted with the ionised gas (plasma) layer that exists outside our breathable atmosphere.

Professor Kristian Birkeland
Norwegian 200 Kroner Banknote

In science, however, times had changed from the openly cooperative days, and so Birkeland attracted some critics, especially from within theoretical physics and mathematics. Nevertheless, he stuck to his guns and on every occasion that called for it got his hands dirty by attempting to prove what he viewed as good science and sound logic. He preferred this practical approach rather than sitting behind a desk in a warm office theorising, like some of his detractors were happy doing.

Birkeland took his theory of auroras initiated by the Sun's power to another level when in his laboratory he built a miniature model of the Earth with an electromagnet inside it and placed that complete assembly within an enclosure that could be filled with a powerful electric field. When he switched the apparatus on, his mini-Earth, or Terrella as it was known, produced glowing rings around both its north and south poles, just like we see with auroras at the poles on Earth, and which interestingly, we now have observed on Jupiter, Saturn, Uranus and Neptune as well. He had demonstrated that his theory worked; all that was needed from that point was proof of an electrical charge flowing toward Earth from the Sun.

It was hard in those days to get a fair hearing if your views went against the orthodoxy of the science establishment. Lord Kelvin (William Thomson 1824 - 1907) seemed to be one particular hurdle in this respect. Although his own contribution as a respected member of the science establishment at that time was great, he did have a bit of a narrow attitude towards the work of some others. It seems also that his thinking was in line with other powerful voices in the science community back then, so altogether, this situation presented a hard nut for any scientist outside that clique to crack. Kelvin is remembered for expressing his lack of belief that electrical currents could exist in space; criticising the idea of X-rays and saying that aircraft would never catch on. He is also recorded as saying that "*There is nothing new to be discovered in physics now.*" This level of hubris typically drove the type of reaction Birkeland was up against in his attempts to have his work receive a fair hearing. In the end, the mainstream science establishment steered its own course based on pet theories that did not include Birkeland's work. However, despite the hostility shown towards his work, it was Birkeland's well constructed equipment and experiments, plus his hard work and detailed analysis of data and an inevitable build-up of supporting evidence that eventually proved his theories correct within plasma science circles. His work was seen as proof of how auroras are formed, but did any of his detractors eat their hats? I think not.

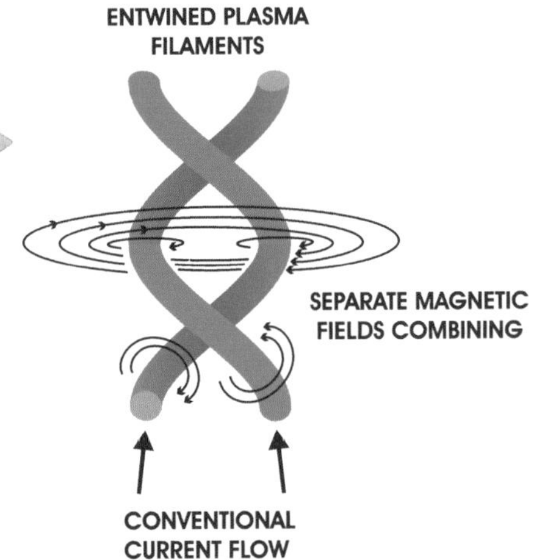


The concept of a helical Birkeland current @ author

A significant discovery that was named in honour of Birkeland's work is where electric currents are observed to flow in filamentary (thread-like) form within plasma. This came mainly from the work of Hannes Alfvén who found that these filaments tended to form in pairs and twist around each other. The resulting helical structure, now known as a 'Birkeland current', is constrained within the combined magnetic field those current-carrying filaments produce. This was extremely important for plasma science because it is the fundamental method by which powerful electric currents flow throughout space.

Credit: ESA & Digitized Sky Survey (Caltech)

As an example from our everyday world, it is Birkeland currents that provide the dancing display inside the 'Plasma Ball' ornaments you can buy from novelty shops. However, on a very much larger scale, these current-carrying plasma filaments feature prominently in the information to come. Here in this picture is an example of plasma filaments as seen in space in the Cygnus Loop supernova remnant.

Birkeland's name will be remembered for many reasons, despite his work and ideas being ignored by narrow-minded people. As a scientist of achievement he came close to being awarded the Nobel Prize at the time of his death, but unfortunately, this did not happen in time. However, two things were eventually done to honour and remember him; a crater on the Moon was named after him and his image appeared on the Norwegian 200 Kroner bank note.

## Dr Irving Langmuir (1881 – 1957) [5-2]

Irving Langmuir

A highly successful American scientist who worked in chemistry then later in electrical and plasma science. Here is a man who among other things, was responsible for discovering the important 'Double Layer' (DL) effect in plasma.

This DL feature becomes apparent when an electric current flows between a charged body and the surrounding plasma. An 'isolating double layer barrier' is formed that insulates the charged body from that plasma. It is sometimes called a "Langmuir sheath." The DL is like the internal structure of the capacitor I mentioned in chapter four, where two electric charges are separated by an insulator. The DL is a fundamental and important concept for plasma science and it is one that has great relevance within the theories I go on to describe. It was also Langmuir who coined the term 'Plasma', probably as his way of representing what he knew the behaviour of blood to be in our own bodies as it forms a protective barrier to defend against foreign bodies. He would have been in a unique position to draw this analogy due to his previous work in field of chemistry.

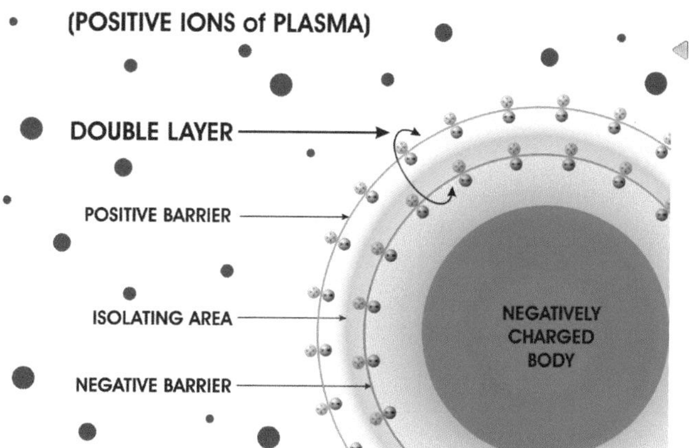

The arrangement of a Double Layer @ author

Another conclusion he drew because of his experience in chemistry and electrical science was that plasma could not be understood or treated under the same rules as apply to normal gases. The tendency to do this had previously been the standard approach of uninformed science, so wrong assumptions had been made about the special role plasma actually plays. This shows up science's understanding of plasma at the time, which at best, seems to have been rudimentary.

Langmuir also invented what came to be known as the 'Langmuir Probe'. This is a special voltage measuring device that can be inserted into plasma without the probe itself being affected by the formation of a double layer around it. The Langmuir probe is therefore capable of obtaining differential voltage measurements from within plasma fields. This is a very important ability that scientists continue to rely on today, especially at NASA. Langmuir was also responsible for a number of other inventions, including the electronic vacuum tube, long-life tungsten filament bulbs and gas welding. He was honoured with many notable accolades from the science community and was awarded the Nobel Prize for Chemistry in 1932.

**Professor Hannes Alfvén (1908 – 1995)** [5-3]

This is the man who is now regarded as the father of Space Plasma Physics. He started as a teenager on his career of scientific discovery in his native Sweden to eventually achieve significant respect from within the global science community.

Hannes Alfvén
Credit: Welinder Jaeger Bergne

In his youth Alfvén set the direction of his life by reading books on astronomy and learning the ropes of electricity and electronics at his local radio club where he built his first radio receiver. His deep interest in radio and electromagnetism was the basis for the doctorate he earned in 1934 for which he wrote a thesis on extremely high frequency radio waves. Throughout the career that followed, his scientific contributions were many in the areas of electrical and plasma science, magnetohydrodynamics, and cosmology in general. But it was for his work in the area of magnetohydrodynamics, the study of induced electric current flow in ionised liquids and gases, that he was awarded the Nobel Prize in 1970. Alfvén's work essentially carried on from that of Kristian Birkeland, but because of his own particular style and the fact that many of his theories directly contradicted the orthodox science of the day, he often found that getting his work published was a nigh-on impossible task to achieve. Alfvén also lamented the direction that science was apparently going in, by commenting on things such as the dilution of professional behaviour in scientific research, the financial distortions that good research had benefited through and then ironically suffered from, and the money-led entity in general that the science establishment had become. In 1986 he said:

*"We should remember that there was once a discipline called Natural Philosophy. Unfortunately, this discipline seems not to exist today. It has been renamed science, but science of today is in danger of losing much of the Natural Philosophy aspect. Scientists tend to resist interdisciplinary inquiries into their own territory. In many instances, such parochialism is founded on the fear that intrusion from other disciplines would compete unfairly for limited financial resources and thus diminish their own opportunity for research."*

Here are some of Alfvén's other contributions. He explained the Van Allen radiation belt that surrounds the Earth. He provided an explanation as to why the intensity of the Earth's magnetic field was affected when it experienced the effects of the Sun's magnetic storms. He put forward a theory for the way in which our solar system was formed, and he explained the mechanism through which the tails of comets are formed. He also explained much about the behaviour of our Milky Way galaxy and that of the general cosmic environment. All of these explanations were based around the behaviour of current-carrying plasma in space.

Another man whom we shall touch on shortly, Dr Anthony L. Peratt, wrote of Alfvén:

[Alfvén made the] *"... novel suggestion that the galaxy contained a large-scale magnetic field and that the cosmic rays moved in spiral orbits within the galaxy, owing to the forces exerted by the magnetic field. He argued that there could be a magnetic field pervading the entire galaxy if plasma was spread throughout the galaxy. This plasma could carry the electrical currents that would then create the galactic magnetic field."*

The science establishment was not ready for Alfvén's views or his style, therefore, many chances to advance our knowledge of how the universe works were lost. He really was a man ahead of his time.

**Immanuel Velikovsky (1895 – 1979)** [5-4]

This Russian author, historian and psychiatrist caused great controversy through his work which suggested an alternative interpretation of ancient history to that previously accepted by most as fact. He had a special focus in this respect on major physical events in the past that had seemingly affected the Earth globally and that were interpreted and recorded in similar ways in those times by ancient civilisations everywhere. These events were apparently frightful heavenly displays and geological disruptions that manifested themselves through powerful forces, including those of an electrical nature emanating from the skies. He suggested these events were responsible for global mass extinctions and he opposed the notion of Darwin's model of evolution. He linked the histories of Jupiter, Saturn, Venus and Mars with that of the Earth in a way that could apparently account for the catastrophes recorded by the ancients, but his theories drew such a different picture that many just could not accept what he proposed. Although popular with the public, his challenging theoretical analysis seemed to make no impact at all on the astro-science or archaeological communities.

Immanuel Velikovsky
Credit: Frederic Juneman

Nonetheless, Velikovsky had and still has his supporters and his work has influenced aspects of the study and analysis of electrical activity in the heavens. Many researchers of ancient historical events say that he uncovered some very interesting truths about the history of the Earth and the Solar System.

**Dr. Charles E R Bruce (1902 – 1979)** [5-5]

Charles E R Bruce

A Scotsman who was an astrophysicist and writer, born near Glasgow and educated at Edinburgh University; Dr. Bruce was to become known as a leading expert on the causes and effects of lightning. His insights allowed him to link what we observe regarding lightning behaviour here on Earth with the activities he had observed at the photosphere 'surface' of the Sun, which by implication, would also apply to the photospheres of all other stars of the same type. Bruce used his scientific awareness of lightning within his considerations of the action taking place in the Sun's photosphere because he viewed it as also being electrical in nature. From the credible scientific position he established through this approach, he was able to provide answers to puzzling questions that had arisen from observations of the Sun; questions that standard astronomy had previously not been able to answer.

His work in this respect included the observed variability of star energy, as indicated by changes in the 'radiant brightness' that some of them display. In addressing this, he believed that the Sun's surface was being subjected to the equivalent of lightning strikes here on Earth, with the exception that these would be on a much larger scale of power, number and frequency of occurrence. From this, he went on to suggest that the variability of star brightness could be explained through an association with a variation in the electric driving force that a star experiences on its plasmasphere from its surrounding cosmic environment.

Bruce expanded this view to encompass electrical events in the more distant reaches of space, notably where Novae were concerned. He suggested that a nova event was electrical in nature and that it did not necessarily represent the death of a star. This was in stark opposition to the standard idea of gravitational collapse producing a nova event when a star dies. Although Bruce adhered to standard theory regarding the non-proven thermonuclear model for star operation, his preference was always for tangible proof, especially where it could be obtained through observation and experiment. Through his core adherence to a pragmatic approach, and in the eyes of many of his colleagues, the output of his research activities was viewed as having come from sound practical foundations. Significantly, he continued in this vein to suggest the existence of Quasars, and that these were born from galactic scale electrical discharges.

Thinking as he did, Bruce was the man who introduced the word 'electricity' into astronomy. This, however, spelled disaster for him as his theories began to spread and eventually become ignored by the mainstream. The day is coming, however, when Dr. Bruce will be vindicated and recognised as the visionary man he was.

### Ralph E Juergens (1924 – 1979) [5-6]

Ralph E Juergens

A Civil Engineer and Science Editor, this American stalwart of the electrical and plasma science community really made his mark with his ideas about our Sun's operation being based on electrical action rather than the story that had evolved since the early 1900s of it being a self-controlling thermonuclear explosion. He was a follower of the work of Immanuel Velikovsky and most of his contributions to electrical and plasma science were to take place after he retired to study Velikovsky's work in detail. Juergens was the person who started talking about the stars, planets, moons and comets, in fact every body in space, as having their own electrical charge. He came up with this belief through his view that space is an electric domain, within which all these bodies reside and interact. He also presented the theory that the Sun and all other stars are actually focal points for electric currents that circulate within galaxies; these currents flowing from an even larger and more powerful inter-galactic circuit. Many who were aware of his work were initially sceptical, but these days we find Juergen's ideas forming one of the foundation pillars of the EU model. Juergens would no doubt have contributed much more, but sadly, he died young of a heart attack in 1979.

### Dr Halton Arp (1927 – present) [5-7]

Halton (Chip) Arp

An American astronomer who worked as an assistant for 28 years to the famous Edwin Hubble at the Mt Palomar Observatory. During that time Arp produced his famous catalogue of 'Peculiar Galaxies', these being galaxies that appear to be physically unusual. In the context of this book, Arp is best known for his views on a particular property of the light that comes from distant objects in the cosmos, this being the 'redshift' I previously explained. From Arp's website biography:

"*Arp discovered, from photographs and spectra with the big telescopes, that many pairs of quasars (quasi-stellar objects) which have extremely high redshift z values (and are therefore thought to be receding from us very rapidly - and thus must be located at a great distance from us) are physically connected to galaxies that have low redshift and are known to be relatively close by. Because of Arp's observations, the assumption that high red shift objects have to be very far away - on which the Big Bang theory and all of "accepted cosmology" is based -* ***has to be fundamentally re-examined!***"

There are now many items of photographic and derived data evidence that support Arp's claim that many objects, supposedly far away from us, are in fact physically joined to objects that are relatively near to us. This directly contradicts what standard science has hailed for decades is proof through the analysis of redshifted light, that our universe started in a Big Bang event and that it still continues to expand. It seems that orthodox astro-science would just not take this convincing and potentially embarrassing evidence on board. In fact, such was the reaction against Arp's revelation that he was denied observing time on all major telescopes in the US and his working life made so miserable that he was forced to leave his job at the Mount Palomar observatory. In the end, and due to not being able to satisfactorily pursue his work in astronomy in the US, Arp moved to Germany where he now works at the Max Planck Institute for Astrophysics in Garching.

What Arp had discovered was that many bodies in space, calculated by the application of redshift theory to be both very close and very far away from us, are actually joined together. How could this be if the application of redshift for distance calculation was correct to do? Could it really be that the Big Bang event never happened and all the other stuff about a 13.7 billion year old expanding universe, is wrong? The evidence does indeed indicate this is the case!

### Dr Anthony L Peratt (? – present) [5-8]

Probably today's leading plasma research scientist, Dr Peratt is well known in electrical and plasma science circles. He has conducted experiments that have provided significant proof of the behaviour of plasma in space and has shown that it can hold stupendous levels of electrical energy.

Anthony L Peratt

These levels of energy have been calculated as significant enough to account for the formation of galaxies and stars and for deep space phenomena such as novae, active galaxy cores, quasars and other sources of high-intensity radio, X-ray and gamma ray emissions. He remains a working scientist but has contributed significantly to aspects of EU theory and research. Two of his major achievements have been ... the definition of 'Peratt Instabilities', which are now considered fundamental to plasma science, and his interpretation of the stone-carved petroglyphs left as records by ancient civilisations in their attempts to describe the mysterious and frightening events they observed taking place all around them and in the heavens above their heads. Dr Peratt's work and reputation has brought great credibility to the EU model and to its community of professional proponents. If it were not for the contributions of Peratt and others like him, then building the powerful case we have for the EU model would have been far more difficult.

This has been a look back into history that brings us to today … so, what is going on now? … who are the people that currently hold all this together and who work to drive things forward? Here, I go back to the names I mentioned in my introduction to this chapter; those of David Talbott, Wallace Thornhill and Dr Donald E Scott. These are the men whose work so impressed me and started me on a road of education and amazing discovery in early 2010. I sincerely thank all of them for the fulfilment I have experienced and the new understanding I now have. So what have these chaps been doing? Well, the current major development and education program seems to have really started with David (Dave) Talbott and Wallace (Wal) Thornhill getting together and comparing notes on their complementary areas of lifetime research. From this they ended up agreeing that the links and overlaps between their areas of research were obvious and that a much bigger story was there to be told. This was a number of years ago, and since that time, together they have produced books, articles, videos, websites and so many other ways of getting EU information out. They look for alliances with other forward looking science organisations and they organise and personally attend public conferences worldwide, all to get the EU model out to open-minded folks in astro-science and to the interested public as well. I would hazard a guess that the breadth and depth of their combined personal efforts would be a real problem to try to quantify.

David (Dave) Talbott

**Dave Talbott** [5-9a] is an American mythologist, science author and scholar who lives and works in Beaverton, Oregon, the place of his birth. His university education led him to a lifetime interest in education itself, and he is responsible for establishing organisations in his home state aimed at improving the quality of education for all and at every level. His commitment to social improvement for fellow citizens has been courageous, effective and is highly commendable. Dave is another who found Velikovsky's original work fascinating and who thereafter decided there was more to investigate about aspects of our ancient history that indicated the possibility of some alternative truths. The upshot is that his subsequent work has been, to say the very least, wide-ranging, in-depth and impressively clear and effective. From his developed point of view, Dave has written articles and books as a modern update to what Velikovsky's original work covered. In doing this, he has been the catalyst for publicising the EU model and for igniting today's fire under the whole subject area. His modern views on ancient history and mythology have been fuel for the personal fires of a large body of interested people; some of whom are eminent scholars and scientists who have begun to look for themselves at these subjects from a new perspective. It is especially significant that in terms of this breakthrough with respected people in academia and science, and together with the existing work of the Natural Philosophy Alliance [5-9b], a new and effective interdisciplinary movement in scientific research has been formed.

The other important aspect of Dave's activities has been his coming together with Wallace (Wal) Thornhill. Out of this relationship many products and presentations have emerged that exquisitely describe the roots, concepts and implications of the EU paradigm against the background of today's accepted theories and the ancient historical record of events. The most recent of Dave's educational videos "The Lightning-Scarred Planet Mars" takes an in-depth look at the surface features on Mars that we are able now to visually inspect very closely indeed. This independent analysis of features on the Martian surface highlights from a new perspective and in very clear, logical terms, how well the evidence for the creation of these features fits with the theories that underpin the EU model. My hat is off to you, Dave, for this and all your other work.

**Wal Thornhill** [5-10] is an Australian astronomer, physicist, electronics expert, author and international lecturer who has an impressive wealth of real world experience in the EU theories of modern astro-science. He is a quietly spoken man who exudes integrity and confidence through the vast knowledge he expresses in his words. His are the original insights that built the core of science theory for the EU model as it stands today; a core that is impressive through its depth and breadth and its references to the work of others on whose shoulders he has on occasion, respectfully stood.

Wallace (Wal) Thornhill

Wal is the main-man in terms of the science behind the EU model. He is the doyen and visionary who has brought its scientific theories together in a form now presented as 'common sense science'. His approach to the clear presentation of scientific information shows up through his co-authoring of books on the EU with Dave Talbott. I must also say that the archive of easily digestible articles on Wal's website is the database most accessed by me during my relatively recent EU education. I have been highly impressed by Wal's ability to generate understandable explanations of theory and I am so pleased to have met him in person on two occasions. These were in London when he was giving one of his world-class lectures on EU theory for the Society of Interdisciplinary Studies in July 2010 and then in the US in July 2011 when he was presenting the John Chappell memorial lecture at Maryland University for the Natural Philosophy Alliance's 18th annual conference.

Born in Melbourne, Australia, his education there took him through university into a globe-trotting job for a number of years, first with IBM and then the Australian Government. He is yet another man who has been hooked by Velikovsky's work. It was from the learning and direction that Velikovsky's ideas provided that Wal decided to take up the challenge of revealing to the world the electric story of our universe. This self-imposed task has now become Wal's lifetime commitment and he is another man whom I respect greatly.

**Donald E Scott**. [5-11] is the third man of great importance to me and to the theories that underpin the EU model. Don is a retired Professor of Electrical Engineering from the University of Massachusetts/Amherst, whose no-nonsense approach I especially identify with through a small kinship I feel because of us both coming from the world of electrical engineering. The obvious difference being, of course, that he taught in a university for 39 years while I would have been fixing motors, switchgear and fuses during that same period! Don's book about the EU, 'The Electric Sky', was the first one I read and the one from which I rapidly formed my first impressions, then a basic understanding of the radical new model being proposed. Thank you most sincerely, Don.

Donald (Don) E Scott

I remember and admire the bravery he displayed not so many years ago by standing in front of a gathering of NASA scientists to present to them aspects of electrical science theory from the EU model. I truly wonder what impression he made on that gathering. Don's work, like Wal's, has been at the scientific roots of EU theory. My own view is that Don has been like "a dug wae a bone" (a dog with a bone) as we say here in Scotland, when it comes to working out the intricate details of mathematical proofs that support findings from the electrical and plasma research laboratories. Anyone who says there is no substantial mathematical support for EU theories should look at Don's and Wal's work and the work of the many other people who have specifically addressed this question.

I titled this chapter "*The work of the honourable but ignored*" … and I fully stand behind the sentiment carried in those words. Most of these people have been, whether some of them got to know it or not, the architects of the EU model we have today. Some of them remain as the standard-bearers whose continuing efforts are now making significant inroads with the current astro-science establishment; they are to be applauded for their efforts and dedication. I must say, however, that I feel quite sad when I consider that it is the classic failings of humans that appear to have been responsible for the suppression of the good work of these and other good people. But on a personal and positive note, the thing that has to a significant extent arisen from this particular aspect, has been my motivation to write this book.

There are many other individuals who are currently doing notable work to help support and develop aspects of the EU model and who absolutely deserve a mention here as well. My hat is off to all those folks as well.

---

We now have all the required ingredients for tackling what I have been banging on about with my references to the Electric Universe model and its theories. Be prepared because there will be a potential revelation on almost every page.

# 6 | The Electric Universe answers I see

This chapter presents my interpretation of the theories of the Electric Universe model. It has been my own understanding of scientific theory and electrical engineering that has allowed me to see the common sense and logic contained in that model. The picture I now have goes beyond notions of gravity's effects to provide me with what I now consider to be the best foundation for understanding the real makeup and operation of our universe. My desire is not to debate this with those who need mathematical or scientific proofs, for these aspects are well covered by those from whose work I have drawn together my thoughts and interpretations. What I present is one lay person's view of what really seems to be going on, and I will leave the defence of the supporting detail to those science workers and other well educated supporters of the EU model who are far better equipped than I am to explain that detail. What I intend as a thinking non-professional, is to point out what I consider to be the stark common sense and logical differences between the questionable 'facts' the mainstream has so far presented us with, and what the more credible EU model reveals and explains through its theories. The final judgement on this I am content to leave to you, the reader.

The Standard Model of the universe invokes the force of gravity and the thermonuclear fusion process to explain just about everything that is physical in the universe. Both these aspects remain unproven in terms of what their overall effects are said to be responsible for, but still the ideas around them remain as dominant 'scientific truths' that people just accept as true. In contrast, the EU model does not claim anything other than to provide alternative theories for actual observations and recorded events for which gravity, fusion and mechanical interactions between matter have traditionally been assigned as explanations. Achieving this more realistic goal is attempted by emphasising the relevance of four particular things; fundamental proven science, the past and present research work of honourable scientists and forward thinkers, evidence arising from the continually improving data we are receiving from space and the application of common sense and logical analysis.

We will see shortly that the case for the real force behind everything being the electric force is a strong one. Electric currents flow and interact within our universe through a network of plasma filaments and sheets of all dimensions and forms. On the grandest scale, the filaments stretch for billions of light years; they are the evolved networks of cosmic scale Birkeland currents that originally drew matter together in the distant past to form the galaxies themselves. Galaxies too have their internal filamentary networks and plasma sheets as sources and conduits for power that sparks the birth of new stars of all types and sizes by initiating forces that draw together available surrounding matter. After a star's birth, when the energy of the Birkeland currents from which it hailed stabilises, its glow is maintained by electric power flowing in from its cosmic environment. The various scales, power outputs and appearances these resulting stars present allows them to be mistakenly interpreted as different cosmic objects that require separate explanations for their creation and operation.

Many of these 'star objects' appear as they do because they are being electrically over-stressed at various points in their long existences: some are in the process of giving birth to planets and moons that are destined to become their own close companions or the companions of other bodies, and some are going through other electrically managed, cataclysmic events. All these bodies are suspended in a sea of plasma so immense and filled with electric power that we cannot begin to comprehend it. Nonetheless, we will start our own journey into this Electric Universe by first considering the universe and its galactic scale filaments of plasma.

### The Environment of the Universe

If we accept that the fundamental laws of physics have never changed, then all the rules we know about plasma, electricity and magnetism will have applied for as long as we care to consider and in all circumstances. This means that from the EU model point of view, there has been ample time and opportunity for electrical currents throughout space to interact with all forms of matter to provide the apparent silent equilibrium we believe we are witness to these days. When and how this all started I firmly believe we cannot safely guess. The universe may be a 'timeless place' and I am glad that EU theory does not attempt to guess at such things.

Ralph Juergens described the universe as an 'electrified fabric' in which charged bodies are immersed. This is important, because the bodies we have talked about; the galaxies, stars, planets, moons, comets, asteroids, meteors, dust and gas, apparently all carry their own electrical charge within this fabric. It is on this basis that all these things interact electrically with this fabric and with each other. It is the Birkeland currents that flow throughout this fabric that mould and spin galaxies and go on to govern their individual and group behaviour and that of the stars within them. We can only begin to imagine how these immense conductors of electric power have interacted in the past to give us the structures and objects we now see when we look into space.

**Galaxy Formation:** We already know that plasma filaments combine in pairs as Birkeland currents (BCs) to transmit electricity efficiently [6-1]. If two BCs meet and the current density that subsequently flows between them is great enough, then an electromagnetic 'instability point' can form. This produces a strong compression force concentrated at the meeting point, which in turn produces a powerful long-range attractive EM force that draws matter toward it. This process is called a plasma 'pinch' effect [6-2].

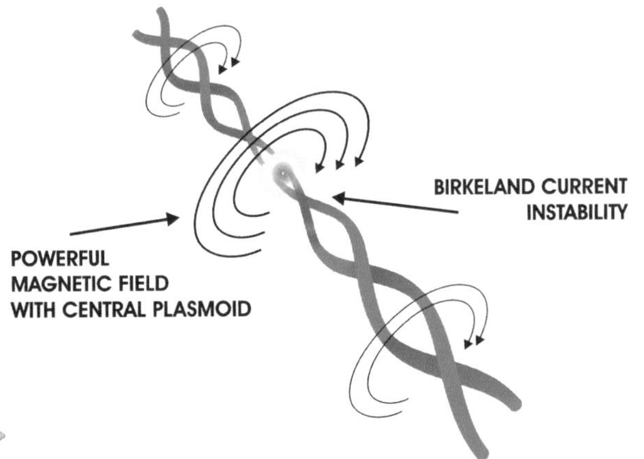


Formation of a Z-pinch event within a Birkeland current © author

This pinch process is better known as a 'Z-pinch' or 'Bennet pinch' event. The Z-pinch is a phenomenon that forms a concentrated plasma structure called a 'plasmoid'. This structure has natural spin and can be pictured mentally, as if you were looking down on the centre of a whirlpool of fantastic energy as it draws everything from its surroundings into it from all directions.

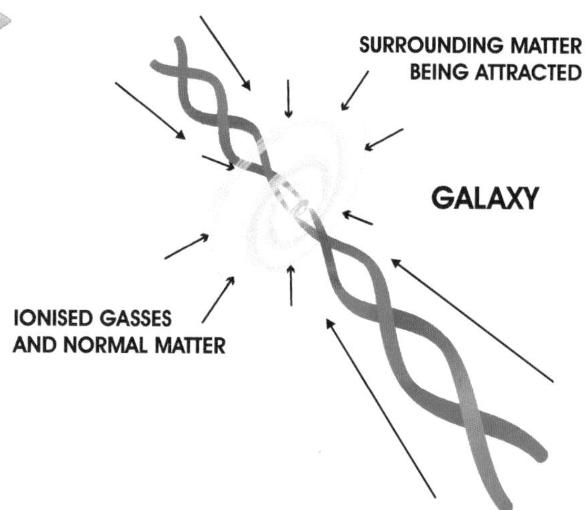


A galaxy formed by a Z-pinch event © author

Think of this on a galactic scale where a Z-pinch event takes place at the meeting point of two BCs to form a spinning plasmoid. This energy whirlpool draws towards it every form of ionised and normal matter from across distances of hundreds of thousands of light years. The result is that all the matter required to form a galaxy is eventually brought together, with the greatest amount of energy and electromagnetic (EM) influence being concentrated at the centre of the mass. Thereafter, a form of 'maintenance current' from the original BCs continues to flow into the central plasmoid. It is the powerful EM force that dominates in the association between all the collected matter so it appears as a solid spinning disk. This is why galaxies hold together as complete structures and it is the basis on which natural rotation can be explained. These are the galaxy formation and operation questions that the gravity model has unsuccessfully tried to explain [6-2].

**Star Formation:** In and around the main bodies of galaxies, smaller BC networks go through the same kind of process, this time in the formation of stars. Eventually, through uncountable instances of this process, most of the available plasma and normal matter will have been drawn together leaving great expanses between new stars which contain only sparsely distributed ionised particles of dust and gas. As mentioned previously, we see stars of all sizes, colours, brightnesses and apparent behaviours, and here is where these differences can be accounted for as we consider the variable aspects of star formation.

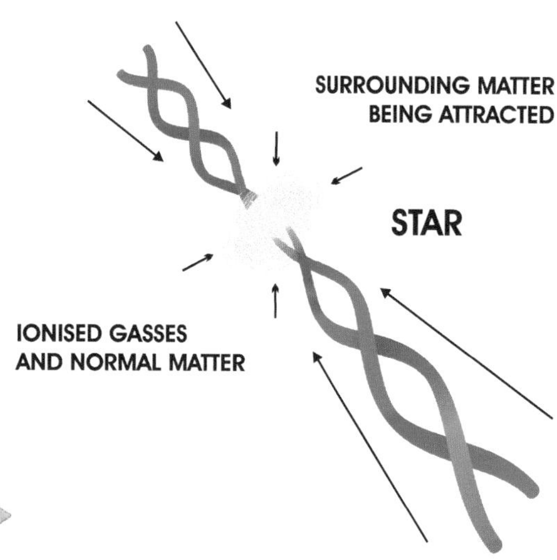


A star formed with others like pearls on a string © author

These variables are: regions of vastly different densities of plasma from which matter is drawn; the various elements that make up that plasma, and the electrical power available for initiating the star birth process. Combinations of these, together with the idea that any new star would have its own sequence of stages to go through over great swathes of time thereafter, is the broad basis on which explanations can be provided for the range of star types and their apparent behaviours that we observe. There are supergiant, giant and dwarf stars, powerful stars, feeble stars and other objects in-between, all of which have associations with the list of 'star type phenomena' that the gravity model has imaginatively produced [6-3].

Over the tremendous amount of time that galaxy and star formation events have taken place, the structure of the universe's power distribution network has evolved. At every scale of electric power that flows into galaxies and stars, this network remains to ensure the charged environment on which their continued operation depends; this is how galaxies and stars are kept 'alive and operational'. No doubt this power grid has changed over time but again, we must remember that our view of the universe is insignificant in terms of the actual time that has gone by. In the very unlikely event of the power supply to a star from its environment being cut, then it would rapidly turn off, just like an electric light would do if we operated its on/off switch. We have never seen this happen and we probably never will, but we have undoubtedly been witness to other events caused by variations in the power flowing into galaxies and stars.

Supporters of the EU model have been aware for many decades of the existence and particular importance of Birkeland currents. However, it is only quite recently that astronomers have been able to detect and image these structures on galactic and stellar scales.

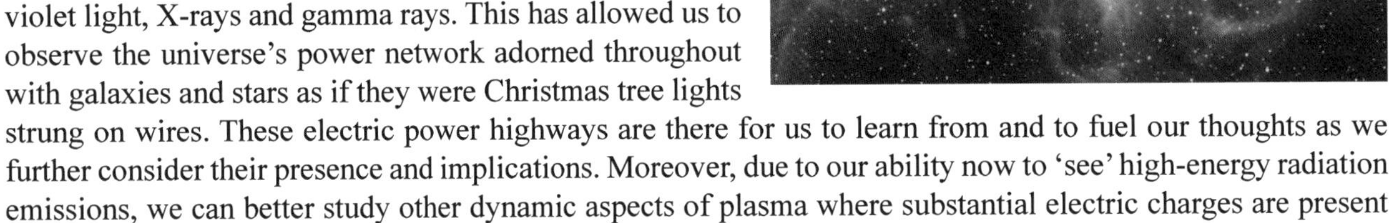

The Trifid Nebula - Courtesy: NASA/JPL-Caltech

We now have excellent images of plasma filaments through the fact that their inherent magnetic fields interact with charged particles in their environment to generate vast amounts of radiation such as radio waves, visible and ultra-violet light, X-rays and gamma rays. This has allowed us to observe the universe's power network adorned throughout with galaxies and stars as if they were Christmas tree lights strung on wires. These electric power highways are there for us to learn from and to fuel our thoughts as we further consider their presence and implications. Moreover, due to our ability now to 'see' high-energy radiation emissions, we can better study other dynamic aspects of plasma where substantial electric charges are present and different, such as in the vast clouds of ionised dust and gas we call Nebulae [6-4].

Given the very reasonable case now established for powerful differentially charged plasma existing throughout space, if there ever could be good reason based on good science and visual evidence to pay serious attention to an electric explanation of galaxy and star formation and operation, then this, I suggest, must be it.

**Stars, Their Combinations, Planets and Moons:** As time passed after these almighty birthing events and as galaxies with their stars grew older and moved around, the vast regions of ionised matter left thinly spread between stars would have facilitated further electrical events that will still apply today. Here, due to the undoubted ongoing cycles of build-up and release of electric charge within and between these regions and with the motion of stars themselves, the overall effect would be to bring about periods of electrical instability between individual stars and between stars and their immediate surroundings. On occasion, stars experiencing this would be forced to take on an excess of charge from their electric environment. This would occur due to high densities of current impinging directly on a star's photosphere 'surface', where that density would be subject to change based on the dynamic nature of the other electrical interactions going on in the star's wider environment. Individual stars could therefore conceivably be forced to take on levels of charge that their mass could not comfortably cope with. This being the case, an electrical stabilisation process would be required that would involve 'shedding' some of this accumulated charge in some way.

It seems to me this type of activity is one that just had to have gone on. It likely still does and probably always will. Much of the reasoning behind this 'electric stress shedding' process goes back to a star's material makeup and physical size, which for now, we will think of as a large ball of gas in the plasma state that has a much smaller solid matter core. These are the two main things that would dictate the level of electrical charge (stress) that stars of all types and gas giant planets could comfortably live with in their own environments. If electrically over-stressed, stars will naturally seek a lower stress situation by dividing into parts or by shedding an amount of their plasma gas or solid matter into space. To help picture the concept of dividing into parts, imagine gently pouring a small amount of water on to a hotplate. If the hotplate is cold, the water will tend to stay together, but if the hotplate is very hot, then the water will quickly be forced to separate into smaller blobs as it absorbs heat energy from the hotplate. This is obviously not a perfect analogy but it provides a useful mental picture.

Most of the stars we see are not just one object, they actually consist of two or more related objects, so perhaps they should really be thought of as 'star systems'. This is a well-known and accepted fact - estimates exist that up to 80% of all stars have at least one partner that is often so dim it cannot be seen. The electrical stress reduction process just described tells us that stars can split into two or actually even more parts in order to reduce the electrical stress they experience [6-5]. This makes basic electrical sense and it is a process that can be reproduced in the electrical laboratory where experiments involving the formation and manipulation of balls of plasma lightning have been carried out.

If a straight split of a star is achieved that ends in two similar parts, the result would be called a 'binary pair'. The electrical stress on the total area of the two new surface areas combined would actually be less than the stress experienced previously on the surface of the original larger body. The two smaller stars may then end up being physically close and orbiting rapidly around each other, just as has been observed with many examples of binary star pairs found over recent decades.

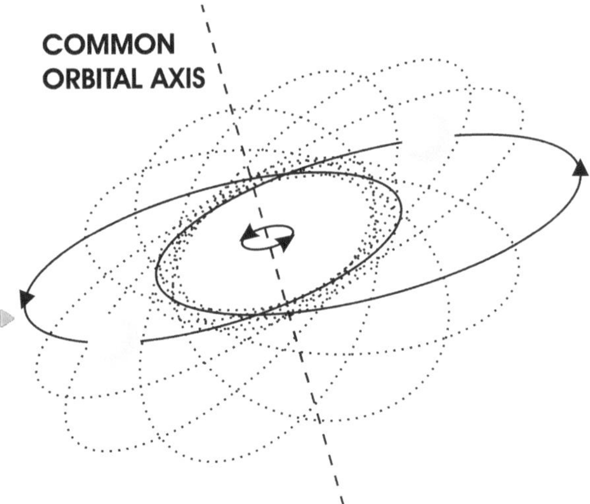


A possible orbital relationship for a binary pair © author

Binary pairs are only one possible result. It is important to note that other configurations can also result due to a greater number of bodies being produced and where some of those bodies may not remain close to the others. Further to this, the electric stress shedding process can produce objects of different sizes, where the original star can end up with what would be judged as a supergiant or giant gas planet orbiting close to it.

The idea of a gas giant planet and its parent star © author

I will mention more on the production of solid planets and moons shortly, but for now, I will provide this summary. A star which is under extreme electrical stress can go through a process of shedding some amount of its plasma from its equatorial region as a gas giant planet. It is also possible that stars and giant gas planets under stress can eject relatively small amounts of solid core material to form planets and moons. All of these gas and solid bodies are born violently and within a short space of time.

This is obviously a picture of events that is in stark opposition to the idea of rocks, dust and gas coming together and fusing into one mass over great swaths of time through a gravity-managed process that in relative terms would not be particularly violent [6-6]. Here we go back to the mainstream idea of how the solar system was formed through the coming together of matter within what has been called its 'accretion disk'.

Our solar system's supposed accretion disk is described in the mainstream model as being a dense and flat disk-like structure of enormous size that had its own natural spin. The matter it consisted of was in the form of rocks, dust and gas that had gradually been brought together by gravity from the wider space environment [6-7a]. During this process, the material collected is also said to have been distributed through the disk by gravity so that the heaviest of the elements were attracted towards the centre, an action which left the lighter gas elements nearer to the outer edge of this solar system sized disk.

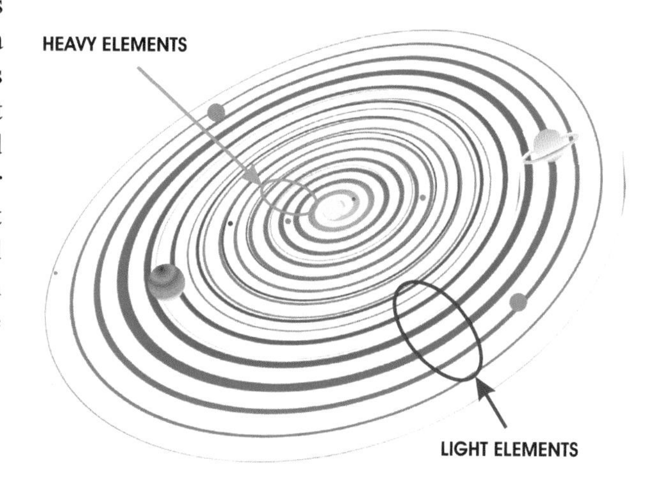


The Solar System disc with heavy matter inside and light matter outside © author

(*Where the natural spin comes from in this description and why the original matter forms a well defined flat disk rather than a three dimensional sphere, are questions that have been speculated on but famously never adequately addressed by the gravity model. The electric model answers these questions.*)

On the other hand, the EU version of how matter in the star creation process is collected together has been covered, in part, by previous mention of the Z-pinch process. A further aspect of influencing heavy and light elements that might help us understand the variety and distribution of planets, as indicated in the graphic above, is a specific electrical process that involves the 'ionisation potential' of different types of elements (i.e. how easily or otherwise the electrons of their atoms can be stripped away so that differentially charged atomic and sub-atomic particles will exist). Due to the greater ionisation potential of heavy elements, they and the denser planetary bodies they comprise are the easiest for electromagnetic forces to influence and to draw towards the inner region of a plasma disk, an effect that would leave in the outer region the lighter gas elements and the larger planetary bodies they constitute. This produces the same overall result in terms of planet distribution as described by the mainstream model, but it is arrived at through considering the effects of the EM force and not gravity. This natural electrical process that apparently sorts elements and planets out is called 'Marklund Convection [6-7b]'. So, given the problems known to exist using the gravity model to explain the distribution of planets in our solar system, could the basic principles of Marklund Convection help us with a basis for explaining why we now have small solid planets (Mercury, Venus, Earth and Mars) orbiting close to the Sun and larger less dense types (Jupiter, Saturn, Uranus and Neptune) orbiting much further away?

There is one other item to highlight about the mainstream belief of how solid and gas planets are distributed in their orbits. Supergiant gas planets have recently been found orbiting very close to other stars in a location where astro-science had always been confident in telling us that only solid planets could be found. Because of this, astro-scientists have been forced to add another poorly explained and ad-hoc idea of 'inward migration' of the gas giants [6-8].

**Star size and power:** The EU model can further provide explanations for the sizes and apparent energy levels of stars that range in dimension from supergiants down to the smallest dwarf type stars. Where current density acting on a star's plasmasphere is not great enough to bring about a high energy plasma mode, this is where we find the red supergiant and red dwarf classes of star. Alternatively, where we see an extremely high current density impinging on a star's plasmasphere, this is where we have highly luminous stars in the plasma arc mode, such as we see with main-sequence stars. To understand this in broad terms it helps to remember from basic theory that plasma has three modes; dark, glow and arc. Here, a variation in voltage pressure will cause a change in current density which in turn will bring about abrupt transitions between these modes. Plasma in arc mode is what brings about the apparent brightness of main-sequence stars, where that brightness and a star's actual 'luminosity' depends on the current density received from its environment.

Despite being a limited summary view of what is actually the case, consider this. High-density current will produce blue and white stars, and low-density current will produce red and dull red stars, with our own 'yellow' Sun being somewhere in the middle between these extremes. This current density relationship with apparent brightness and colour is applicable to all sizes of star, so using this line of thinking we can form a useful view of how big blue supergiants together with big red supergiants and little red dwarfs, appear to us as they do.

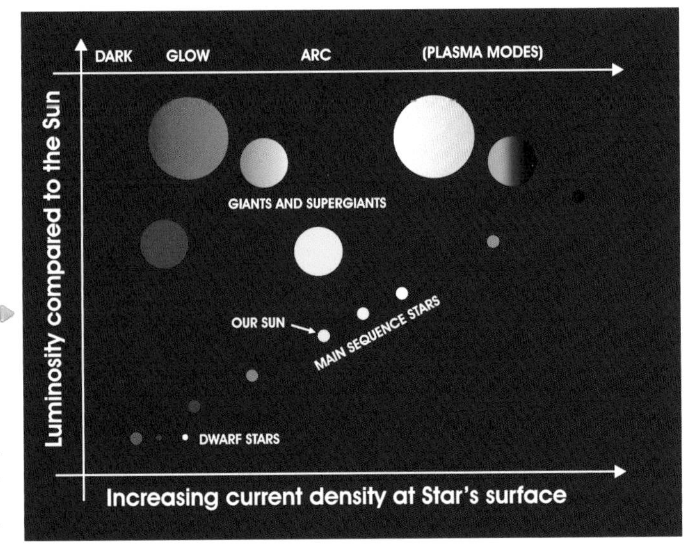


The distribution of star types due to the mass and energy involved (From an original graph by Don Scott) © author

Some stars receive very low current density from their environment so their appearance is that of a dull red glowing ball. These are stars at the low energy end of the scale where we have red, brown and white dwarfs.

Even further down in relation to the thermonuclear star model, this is where we find the so-called 'failed stars' but that term has no relevance in the electric model of a star. It is simply the case that these are stars whose plasmaspheres are not being electrically excited enough to give off a great deal of visible light. In our own solar system today there are bodies that have a relationship with this idea. Jupiter and Saturn both have enormous invisible electromagnetic bubbles around them, as do all other bodies in the solar system, plus very relevantly the solar system itself, but these two gas giants are particularly important because of their possible histories about which I will say more later.

Taking Jupiter as the example, the electromagnetic bubble that surrounds it, also referred to as Jupiter's plasmasphere or magnetosphere, is extremely large and currently not visible because it consists of plasma in dark mode. If the outer surface of that plasmasphere, being Jupiter's electrical barrier to the rest of space, were to receive per unit area a much greater current density than it does at the moment, then it would start to glow and Jupiter would appear to us here on Earth as if it was a much larger brown/red dwarf star. The diameter of Jupiter's plasmasphere is many times its own diameter, and if we could see this in the night sky it would look similar in size to the full moon. Incidentally, if this were to happen, then all of Jupiter's current moons would orbit inside this glowing bubble, so anyone standing on one of those moons would only see above them a visually impenetrable purple glow right across the sky; there would be no day or night, no seasons, no stars to see and the temperature and level of illumination would be very constant. (Associated information [6-9A] [6-9B] )

Here is a contradiction that underlines the relevance of questioning what we are told about stars. There are classes known as T and L type dwarf stars. Relative to normal stars, these are very cool indeed with estimated temperatures of between 600 to 1000K. Interestingly, these are temperatures in the same region as areas on the surface of the planet Venus. Temperatures this low indicate that the thermonuclear fusion process cannot possibly be occurring inside these bodies. Yet X-rays have been detected coming from similarly cool brown dwarf stars, where again, the low temperatures involved are fundamentally incapable of initiating the production of this powerful type of radiation. Straightforward evidence like this that indicates things are not right with the thermonuclear theory of stars should be all that is needed to drive a more open and inclusive investigation, but sadly, it does not.

Mainstream astro-science can talk about size, temperature, colour, radiation emissions and behaviours of their range of star types, but it does not matter, for there seems to be an electrical bottom line to what has traditionally been interpreted as different types of star. According to the EU model, all stars started as different combinations and amounts of ionised matter as plasma, being brought together by the EM force then maintained in appearance as the stars we recognise or as other phenomena by a certain level of current density delivered to them from their environments. The standard list of star types is misleading and the thermonuclear star theory is wrong. What we see in space are fundamentally all the same things; concentrated bodies of plasma reacting to different levels of electrical energy.

**Producing planets and moons:** I mentioned earlier that stars and giant gas planets under exceptional electrical stress relieve some of that stress by shedding into space both portions of their plasma bodies and material from their internal solid cores. Stars will likely do this more during their formative years as the concentrated EM forces that formed them in Z-pinch events subside and as their interaction with other young stars is more likely. It is also possible that long-term fully formed stars and giant gas planets will do this on occasion due to influences from their electrically dynamic environments, forcing them to take on an uncomfortable excess of charge. From this it seems possible that two expulsion methods could operate.

Amounts of relatively light plasma gas material may be ejected from the equatorial region of the parent body into its ecliptic plane, and due to the concentrated forces present at its poles, relatively small amounts of solid material might be ejected from there to become rocky planets and moons. This idea is inclusive so it does not ignore that circumstances may exist where solid bodies could also be ejected from the equatorial region. Given the all-encompassing relevance of electrical and plasma science, I believe that within the grand scales of energies in play and the amounts and types of plasma material involved here, we will eventually find explanations for all the types of bodies we find in space.

Giant gas planet and solid planet expulsion methods and the adoption of rotation and axial tilt © author

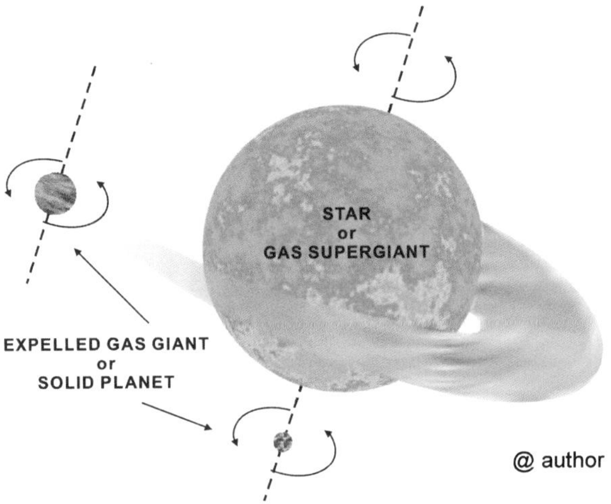

**Natural spin:** It is likely that matter ejected by a star or supergiant gas planet to form smaller gas or solid bodies, will over time, as a satellite body of its parent, tidally adopt the same tilt (angle of inclination) and direction of rotation that its parent body has. This provides a basis from which we can explain possible associations between the gas giant planets, solid planets and moons in our own solar system. (Remember this, for I will return to it shortly.)

There is a further very important aspect of 'body formation' to mention, one that the gravity-centric accretion process cannot explain on any scale whatsoever. This is the question of why galaxies, stars, planets and moons all have their own natural rotation. To provide an answer we must consider what basic electrical theory tells us about magnetic fields around conductors. The natural torque magnetic fields produce due to their circular polarisation around a current-carrying conductor (a current-carrying filament of plasma, or Birkeland current in our case), is a universally accepted fact that is proven by electrical laboratory experimentation. This being the case we are presented with an electrical explanation for natural rotation existing in and around the Birkeland currents of a cosmic electrical circuit. Due to this we can have further confidence in the relevance of EU theory [6-10].

If we accept that stars with their attendant planets and their moons move around in various configurations as independent systems in space and that interactions between these star systems will inevitably have gone on against a scale of time we cannot imagine, then we have a basis on which to consider how the 'electrical capture' of stars, planets and moons can take place. Remember that I previously mentioned a bubble of electric charge existing around the solar system. This will apply to every star system and the bubble itself will extend far beyond any planets that a star system may have. This bubble is given the name 'heliosphere' or more generally, 'astrosphere', and its outer edge in contact with deep space is called the 'heliopause'. Due to the double layer effect also previously mentioned, the heliopause is where the greatest voltage difference will exist between any star system and deep space.

Interacting star systems, one more dominant than the other. (This image does not represent the fuller story) © author

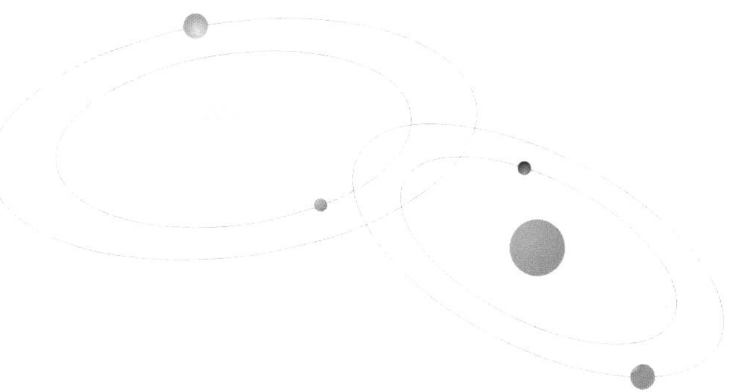

When the heliopauses of two star systems come into contact with each other, the 'action' begins. The star system with greatest level of charge will electrically dominate then manipulate all of the bodies within the less powerfully charged system. The 'touching' of heliopauses brings about an immediate transfer of positive charge from the less dominant to the more dominant system, so the star in the less powerful system will quickly lose its glow and stop looking like a star since it is negatively charged in relation to the star of the dominant system. The less dominant star then becomes a gigantic comet, shedding matter in an effort to regain electrical equilibrium. Then, it and all its original planets are swiftly 'captured' by a charge exchange mechanism within the original dominant star's ecliptic plane. That same charge exchange mechanism will further manipulate all the new arrivals into electrically and gravitationally stable orbital locations within the new star system. Due to its previously large glowing plasmasphere then being 'switched off' the much smaller remnant body of the captured star will find itself treated the same as the planets that once were its own. It is possible, in fact likely, that it would end up having the appearance of a giant gas planet such as our own Jupiter or Saturn, possibly complete with debris rings leftover from the encounter.

The scene is therefore set, so consider our own Sun thousands of years ago with whatever planets it had at the time, and a brown dwarf system with its own planets coming into electrical contact with it. If we apply the above explanation then what we see today is an apparently peaceful snapshot of planets and moons that will have experienced great chaos in the past. An intriguing proposal now surfaces, one that may explain the particularly strange makeup of our solar system. This proposal goes further than the hard-science of electric and plasma cosmology to seriously consider the recorded evidence that exists from ancient times of cataclysmic events of epic proportion that occurred in those days.

Here, it has been suggested that just a few thousand years ago a body we may here call 'proto-Saturn' in the form of a brown dwarf star appeared as a 'second sun' with its own retinue of planets, only for that star system to be captured by our current sun's more powerful system [6-11] [6-12]. To some this will sound far fetched, but evidence exists that suggests this, or something close to it, may have happened. The line of thinking that revealed this evidence from the past has been like uncovering the facts of association that the police fingerprinting process achieves – physical evidence leading to the uncovering of further evidence, then the establishment of facts that lead to conclusions. There also exists evidence from astro-science, one example being the 'axial tilt' of bodies as they rotate in their orbits; in this case, the similar tilt of certain planets in our own solar system. This important aspect stems from the broad acceptance that bodies in space that are in relatively close proximity and which have very similar angles of inclination (axial tilts) as they rotate, are likely to have a fundamental relationship. Well, lo-and-behold, the angles of inclination to the vertical axis of Saturn, Mars, and the Earth are all very close to 26˚. This is significantly different from other planets and the Sun itself which has a tilt of 7˚ to the orbital plane of Earth, otherwise known as the ecliptic. One interpretation arising from this that is supported by the work of very credible researchers, suggests that Saturn was indeed a dwarf star accompanied by at least Mars and Earth, and that this 'star system' was captured complete when it came under the Sun's more powerful EM influence. (In fact, we know there were other bodies because Saturn still has a large number of satellites, including one of planet size with a heavy atmosphere - Titan.) Saturn then lost its vast electrically maintained radiant glow, and along with its planets was re-located to the orbits they now occupy.

Gravity remains unexplained by modern physics. In the EU model, the gravity of a planet is modified strongly by charge exchange, a process which is capable of manipulating charged bodies in an environment that offers no mechanical friction or other hindering forces. Having lost the source of its radiant energy over a relatively short period, the dwarf star we now call Saturn would have assumed the appearance it now has of a gas giant planet. The detail behind this process can also account for the rings of Saturn as being material left over from this energy shedding event. As I said, intriguing stuff, and I realise this idea is a bit of a departure, but it really does have evidence to support its credibility from today's science, geological studies and from the analysis of ancient recorded history and mythological accounts [6-13]. The basis for judging this 'historical evidence' as substantial arises from similarities recently found between petroglyphs carved in stone from all around the world and through stories with common threads handed down within tribes and isolated civilisations from every continent. These were people who could never have communicated with each other to exchange this information, so why is it that so many of their recorded accounts tell the same story? For anyone particularly interested in this developing historical database and its analysis, please look at the work of:

David Talbott via www.thunderbolts.info - Marinus Anthony van der Sluijs at www.mythopedia.info - Dwardu Cardona via www.velikovsky.info/Dwardu_Cardona - Ev Cochrane at www.maverickscience.com

You can also find books by these researchers at www.mikamar.biz/thunderbolts-product.htm

Getting back to our own solar system today, what keeps everything in place? Given that gravity theory is only capable of predicting the interactive behaviour of two bodies and not of systems that contain more than that, it seems there must be some stabilising feedback mechanism that can modify gravity. In the solar system, planets orbit the Sun, moons orbit planets and everything seems to be stable, but remember, this is only a snapshot in terms of the time that has already passed.

Our star is a component within two electric circuits; one interstellar and the other local to the solar system itself. Without exception, all electric and magnetic circuits must be 'complete' in order for current to flow and for magnetic fields to exist. The colossal interstellar maintenance currents that flow into the Sun enter at its north and south poles [6-14]. The other internal solar system circuit has positive protons drifting away from the Sun through the heliosphere to the heliopause and negative electrons drifting from all regions of the heliopause back towards the Sun.

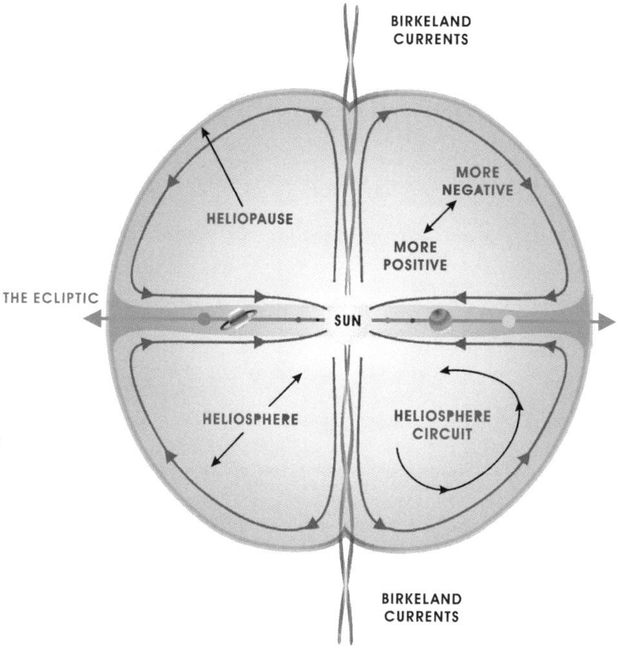

The circuits and the Ecliptic plane within the Heliosphere © author

We must remember that the heliosphere actually is like a sphere, so 'charge drift' is occurring from every direction towards the Sun's plasmasphere at the same time. This, however, is concentrated in some regions more than others, for example in particular along the flat disk-like structure around the sun's equator known as the 'ecliptic'. It is within or close to the ecliptic that the orbits of all the planets are to be found and the disk itself extends out to the solar system's 'heliopause'. Think of the heliopause and the heliosphere together as one body, like a slightly dented globe at top and bottom that would appear like a doughnut if it were flattened more. In electrical terms, the positive Sun would be known as the 'anode' and the more negative heliopause would be known as the 'cathode'. Just as electrons do in a normal electric circuit when they flow from cathode to anode, they will therefore naturally flow from the heliopause towards the positive Sun. If you had a voltage meter with very long test leads and were able to put one lead on the Sun and the other on, say, the Earth, then you would find the Earth to be negative, but not nearly as negative as measuring the voltage at, say, Mars, or at any of the other planets further away. Think then of the planets as bodies that are negatively charged to different extents, floating within this gradient of charged plasma that changes from a positive maximum at the Sun to a negative maximum at the heliopause.

Given what has been said so far, go on to consider the various distances that the planets are from the Sun and the orbital stability they seem to have. The negative charge associated with each of them will have settled, more or less, at a level that matches the mass of the planet within its heliosphere environment and also its relationship to other charged planets that will transit on a regular basis in adjacent orbits. The apparent steady and predictable orbits of planets is therefore an indication that the EU model of an electrical modification to Newton's clockwork mechanism is in operation. How gravity is modified by the electric force is an aspect described under the term EMOND (Electrically MOdified Newtonian Dynamics). Wal Thornhill describes this on his website as the process whereby variations in the electric charge of a body (planet or moon etc.) alter its mass and therefore the gravitational pull that it can exert or be influenced by within its environment [6-15A] [6-15B] [6-16].

The distance from the Sun to the heliopause is estimated to be between 115 and 175 AU or 10,700,000,000 to 16,300,000,000 miles. Beyond the heliopause lies a mind-bogglingly large volume of space that contains extremely sparsely distributed particles of plasma gas and dust. Suspended within that apparent emptiness are countless identical situations of other stars with their astrospheres and astropause boundaries and no doubt their planets and moons inside. But remember, this is just within our own Milky Way galaxy.

We also should remember that within our living memory in our thin biosphere environment here on the surface of the Earth, we humans have been largely isolated from the direct influences of the powerful EM forces that exist in space. This limited experience is a substantial reason why we are not more aware even of the existence of these forces. Our on-going isolated survival is due mainly to the barrier formed by the earth's protective magnetic field, the 'magnetosphere' [6-17]. The magnetosphere acts as Earth's protective shield and provides a path around which high-speed charged particles from the Sun are guided away from the Earth. Some particles that are not diverted do reach Earth at the poles, where they come in contact with atoms of certain gases in the upper layers of our atmosphere.

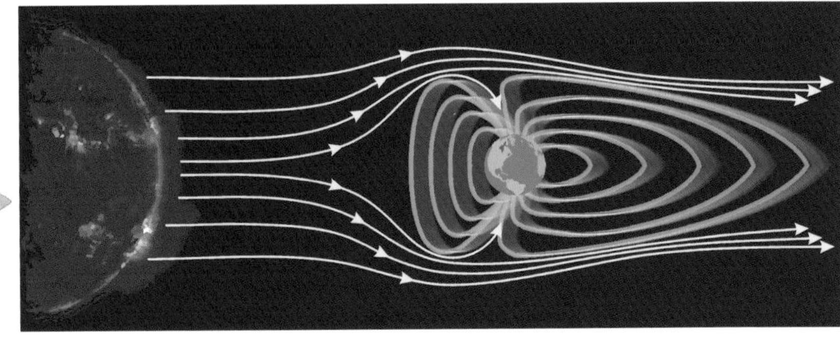

The flow of charged particles along the ecliptic and their interaction with the Earth's magnetic field. © author

The result of these energetic particles striking molecules of oxygen and nitrogen gas in our upper atmosphere is what causes those gases to be ionised and thereby appear to us in plasma glow mode. We see this from the ground in the northern hemisphere as the 'Aurora Borealis' and in the southern hemisphere as the 'Aurora Australis'. Depending on the altitude at which this occurs, we see various colours that appear to us as dancing red, green and violet curtains of light.

One of my lifelong hobbies has been amateur radio, where my practical activities have included making various pieces of equipment, carrying out experiments in communications, and occasionally talking with people in different places around the world. One thing we 'radio hams' are taught early on is that when we use certain 'short wave' radio frequencies (RF) for communication, we should not expect the same frequencies to carry our communications signals effectively at all times of the day. This is the same as saying that certain frequencies work best at different times of the day. This is because the ionised layers that form our upper atmosphere either bounce radio frequencies off them or let them pass straight through and out into space.

This behaviour is due to the interaction of the transmitted EM radio waves with the various densities of plasma that make up those layers. The density of the plasma layers is dependent on the time of day, where varying amounts of radiation will be encountered by those layers in the form of ultra-violet and X-ray emissions from the Sun. These varying densities and their elevation above ground are the deciding factors for which radio signals will be passed through to space and which will be absorbed and/or reflected back to Earth. If the radio signal is higher in frequency than what the density of the ionised plasma will react to, then it will be allowed to pass straight through. If it is lower in frequency, then it will be bounced back to Earth, with some amount of the transmitted RF energy being lost (absorbed) in the process.

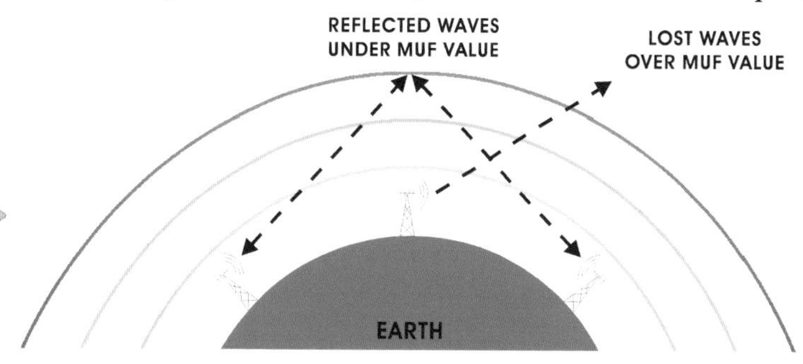


Radio waves are either reflected or passed by the ionised layers of our atmosphere © author

This, in radio terms, defines the Maximum Useable Frequency (MUF) for worldwide shortwave communications at any time of the day. Interestingly, this phenomenon works both ways - Earth to space and space to Earth. This meant that before this effect was known about, certain radio frequencies from space had never been looked for or recognised by us. One radio amateur in particular did fundamental work on this that literally went beyond the MUF barrier to look for radio signals from space. This person was Grote Reber, holder of the US amateur radio call sign W9GFZ. Building on the earlier work of radio engineer Karl Jansky, Reber was likewise convinced that radio waves already existed in space and that they emanated from natural sources. In his first attempt to prove this, he constructed his own parabolic dish antenna at home in his back yard, built his own specialist equipment and carried out experiments that would prove the accuracy of these ideas. His claims and efforts were ignored by the mainstream for ten years until eventually the evidence grew and he was recognised as having been correct all along. Grote Reber was actually the man who began Radio Astronomy through his dedicated work and by producing the first 'radio emission map' of our Milky Way galaxy [6-18].

As development progressed in radio astronomy, due to encouragement from the discovery of a whole spectrum of radio waves by the first space probes with appropriate sensing equipment, an even greater focus was applied to the activity as its truly huge potential began to be appreciated. Previous to all of this, radio waves were not even suspected to exist out there because space was thought by most to be totally empty. This aspect was what Reber had always thought was not the case due to his own science background and his knowledge that our upper ionised atmosphere would have kept certain radio waves from us by blocking or attenuating most of them from being detected by the relatively crude equipment we had in those days. This was what Reber saw as his personal challenge, so, in the spirit of radio amateurs everywhere, he got down to learning more about and proving the correctness of his ideas. The important things to note for our purposes are that radio waves can be blocked by ionised plasma; that plasma surrounding all bodies in space will do precisely the same things that we know it does with Earth and that the properties of these plasma barriers can change depending on the density of charged particles present at any point in time.

I briefly described this electric (electrostatic) barrier when I mentioned in chapter five the man who discovered it, Irving Langmuir, who also gave it the more commonly used label of Double Layer or DL. These DLs will exist between all charged bodies and between those bodies and the plasma environment of space. The presence of DLs as regions where voltages of often enormous value are separated from each other, sets the scene for powerful electrical interaction to take place, should the difference between the voltages involved become extreme. This is the case between the positive plasma environment of the Sun and that of Earth's more negative plasmasphere. It is also the case between the solar plasma and the significant negative charge possessed by comets as they travel in their eccentric journeys around the Sun. (We shall be looking closely at comets shortly.) Think many millions and on a galactic scale, even billions of volts of difference here. DLs are very important indeed, so keep them in mind [6-19].

We learned previously that electric current flow in a conductor forms an associated magnetic field around it. Where this happens with plasma as the conductor, the magnetic field constricts the cross-section of the plasma into filaments that entwine in the pairs we call Birkeland currents [6-20]. The magnetic fields that hold Birkeland currents together cannot be observed but the filaments inside can be detected due to the EM radiation they emit. We said before that plasma is not a perfect conductor; this means that when current flows within a plasma filament some energy will be dissipated (lost) in the form of heat, radio waves and X-ray radiation.

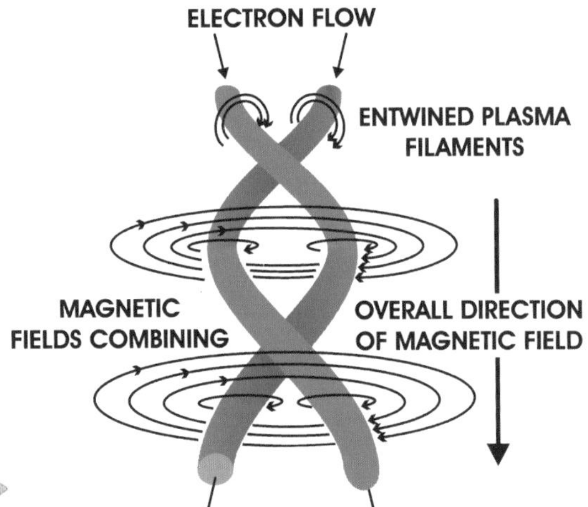


The magnetic forces that keep Birkeland currents together © author

The sensitive equipment we have can detect this radiation so images can then be constructed that let us understand the filamentary nature of structures in space. One of the best sources for imagery has been the Chandra X-ray observatory launched in 1999. The X-ray detectors it carries have revealed to us not only wonderful structures in our own galaxy but many that lie billions of light years further away.

The Cygnus Loop. Credit: ESA & Digitised Sky Survey (Caltech)

What appeared before to us as vast empty spaces are now showing up as regions that have within them filamentary networks that deliver power to galaxies and stars and to vast regions of dust and gas where the formation of new bodies is going on. Appreciating the pervasiveness and power of Birkeland currents on the grandest scales helps us understand how plasma networks have formed the universe and remain to dominate it.

**Stars again:** With more now in place, we will go back to stars. We said they have solid cores and that, depending on the charge level they have, their plasmaspheres will extend sometimes to great distances. The size and colour of what appears to be their outer glowing or arcing surface is therefore no indication of the size of the core, in fact, we have no way of telling much about a star's solid core except that electric theory says it is cooler than the photosphere. A star's appearance is also no indication of its age, as has been assumed for so long by astro-science. Rather, the apparent size and colour tells us about the physical dimensions of the plasmasphere around the core and the density of current flowing onto and into the star at this single point in time [6-21]. In terms of all low-energy stars that do not 'shine' like typical stars, these bodies have until just a few decades ago, been very hard indeed to detect.

A comparison of star and planet sizes to Earth
Credit: NASAJPL-CaltechUCB

The way we do this today, however, is by using special telescopes and detectors that are able to 'see' the radiation that stars and other bodies and events give off. Much of this detection takes place using infrared (heat) radiation, but still more is achieved by looking for other wavelengths as well like radio, ultra-violet, X-ray and gamma ray emissions. (*Noteworthy: Some researchers say that many other stars like our Sun have at least one companion body. Our own Sun is rumoured to have one that has not been observed for thousands of years, so its existence has never been confirmed or in any way acknowledged by modern standards. It has been said that it, along with its own attendant planet(s) are on a long elliptical orbit that sees them passing through the solar system very rarely. This dwarf star has been assigned a variety of names over time such as, Planet X, Marduk, the 12th Planet, Nemesis, The Destroyer, Nibiru, and others. When looked further into there is confusion around the actual configuration of the main star and any planets it may have, but those who study this subject generally support the model of a brown dwarf star having at least one planet. Whatever the facts may be, it has to be said that a search for this brown dwarf has been and is being conducted on the basis that by the law of averages, a body such as this may actually exist, and that its long orbit around the Sun, which has been said could be 3,600 years or more, is the reason why we have no formal record of it.* [More on stars [6-22A] [6-22B]])

It is not just relatively small stars that can have low-power levels; there are much larger stars which also appear not to be operating in arc mode. The volume their plasmaspheres will encompass will be gigantic around what must be their relatively tiny cores. An example of this is the ninth brightest star in the night sky which has been classed as a red giant; this is Betelgeuse in the Orion constellation. Estimates have been made of the diameter of this star and it is typically quoted as being around the diameter of the orbit of Jupiter. Can you imagine a star of almost a billion miles across?

Estimated size of Betelgeuse and its location Credit: NASAJPL-Caltech UCB

The EU model of stars suggests that inside the large dimly glowing plasmaspheres (atmospheres) of brown dwarf stars, the space between their solid core and the inside of their layer of glowing plasma would be the ideal environment for life to thrive. Although much larger and hotter than a brown dwarf, Betelgeuse has been proposed as a candidate for having planets orbiting within its glowing plasma shell. Furthermore, as I mentioned in relation to Jupiter and its moons, it is said that planets in this circumstance would likely experience constant temperature and light exposure and they would not be subject to seasons like we are here on Earth.

The idea of a variety of physical locations across all of which the energy received at all wavelengths is constant, is like placing a thermometer in random locations throughout a sealed and heated enclosed space. No matter where you take measurements a similar amount of heat energy will come from every direction - the bulb of the thermometer in this case representing a planet orbiting inside the plasmasphere of a low-power star [6-23].

However, this situation could be seen as having its drawbacks, for again, as I said previously about Jupiter and its moons, to look up into the sky from the surface of a planet located inside a brown dwarf's plasmasphere, would be to see only a glowing purple haze from horizon to horizon. There would be no stars to see beyond that opaque barrier and the inhabitants of such a planet would have no clue about anything existing beyond it. The ionised bubble of the star itself would be like our own Sun's heliosphere, isolated from space beyond by a charge separating DL that would not allow the easy passage of radio waves and which would perform the same role as our own Sun's heliopause. Because of this, the inhabitants would have no reason to view radio technology as anything other than a tool for communications across the surface of their own planet or possibly with other companion planets of the same star. They would also have no idea of the existence of galaxies or the potential for life to exist beyond the beautifully glowing canopy above their isolated world.

**Nebulae: 'Nebula'** is an old word in astronomy. It was used long ago to describe very dim patches of light in the night sky that were thought only to be clouds of dust in our own Milky Way galaxy. As astronomy progressed and optical equipment improved, some of these were actually found to have structure to them. Most of these turned out to be galaxies of one form or another, far beyond our own, but true enough, some really were just regions of dust and gas within the Milky Way. The name itself has been retained but now only refers to regions of apparently glowing dust and gas. In EU terms, Nebulae are fantastically dimensioned regions of ionised dust and gas as radiation-emitting plasma that is manipulated by powerful EM forces as opposed to puny gravity. Nebulae result from cataclysmic electrically driven events such as exploding electric stars. They can range in size from one to many hundreds of light years across; this is in the order of a few trillion to thousands of trillions of miles. Filamentary Birkeland currents flow within and through them and act there to draw together plasma matter to form new stars through powerful Z-pinch events and to bring about other spectacular electrical interactions. When dwarf star and planet-forming events occur and we observe them here on Earth, we call them 'Nova' events.

The Monoceros Nebula. Credit: ASAJPL-Caltech UCB

Note that **Nova** is the name given to an 'event' rather than to a solid object. The word 'nova' comes from Latin for 'new'. It was used by astronomers long ago as a reference to 'new stars' that suddenly appeared in the heavens. Perhaps confusingly, the terms 'nova' and 'supernova' can be used by today's astro-science to represent both the birth of stars and their gravity-managed death; one has to read the detail to understand the use of these terms that are supposed to represent the differences in the apparent power involved in each.

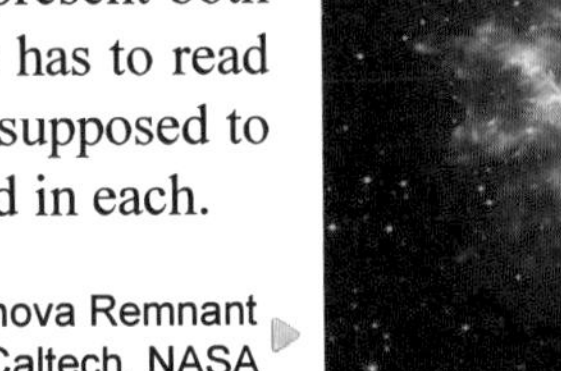

Casseopeia Supernova Remnant
Credit: O. Krause (Steward Obs.) et al., SSC, JPL, Caltech, NASA

The observed amounts of energy and levels of radiation produced by nova and supernova events has always been a problem for astro-scientists to explain, most likely because they only have available to them the inadequate tools of feeble gravity and mechanical shock as a basis on which to meet this challenge.

In EU terms, novae are caused by stellar electric discharges on a massive scale during the birth of dwarf stars, gas giant planets, rocky planets and moons, rather like mega solar mass ejection events [6-24]. Supernova events are cataclysmic releases of energy in cosmic electric circuits that are focussed on Z-pinch events. These pinches can mark both the birth of stars as has previously been described or their destruction through the explosion of stellar scale Double Layers (DLs) that can form around stars that suffer extreme electrical stress from power fluctuations in their cosmic environment. The gas and dust of the often symmetrical glowing remnants of these explosions are what we call planetary nebulae [6-25].

We should now be well aware that powerful discharge events occur when stars become over-stressed and attempt to reduce the charge they are forced to carry. They can also explode completely if the electrical stress imposed on them from their environment is sudden and unmanageable in nature. The energy released by the pinch event and the breakdown of the DL barrier by such an event produces copious EM radiation that interacts with the two axially aligned Birkeland currents that supply the event to give off X-rays that betray the structure of the BCs as a series of concentric 'sheaths' that surround their respective central axes. These BCs having been asked to rapidly carry a vast amount of current also become visible due to their ionised matter switching from dark to glow mode. Images of this type of event, where powerful radiation is generated and detected, now reveal the normally hidden Birkeland currents and allow us to see their bi-polar orientation, as here in this picture of the Ant Nebula.

Ant Nebula - Credit: NASA STScI

The visual result is the traditional hourglass shape that is often seen with supernova events. There are though instances where we will not see this obvious form due to Earth not being oriented in relation to these events with enough of a side-on view for us to see the expanding plasma material around the Birkeland currents. It must be underlined that the energy available from mechanical shock forces cannot explain this symmetrical hourglass shape. In fact, if gravity and mechanical forces were really to be involved everywhere, we should at every turn, instead of an array of geometric shapes and flat structures, see an expanding globe equal in all directions.

When supernovae involve extremely high levels of energy, standard astro-science identifies them as deadly Gamma Ray Burst (GRB) events. EU theory provides strong reason for us to believe that GRBs are just an extension of the process we have already discussed here, this time, however, at a super-tremendous scale of power and violence. The radiation a GRB emits is deadly to organic material; this includes our own human bodies. If a person with no shielding was physically close to a GRB event, its destructive radiation would kill that person instantly. If a GRB occurred a few light years from Earth, the surface of our planet would be completely irradiated and sterilised a few years after that event. All types of novae and GRB events are related. They are the cataclysmic results of electrical discharge events that bring about the breakdown of DL barriers to produce colossal amounts of radiation [6-26].

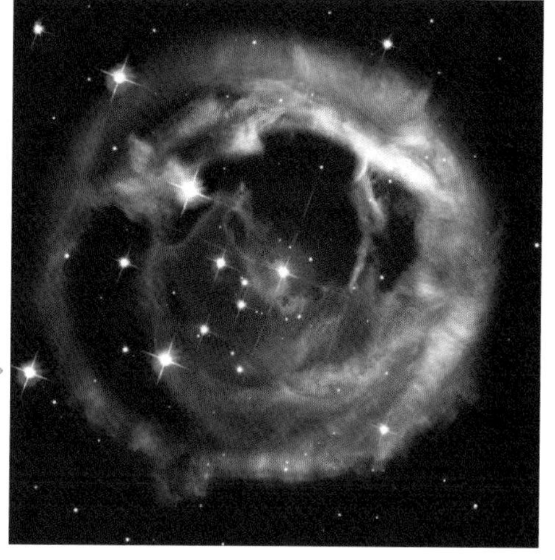

Supernova type event V838 Monocerotis - Credit: NASA - Lynn Barranger STScI

Copious amounts of RF, ultraviolet, X-rays and gamma rays are constantly being produced by powerful events throughout the cosmos. The detection of this radiation is now made possible due to radiation sensors being integrated with standard observing equipment. It seems that naïve assumptions were made previously in regard to observed events and other phenomena because our early observational capabilities were limited only to the visible part of the EM spectrum where our eyes distinguish light and colour. This new ability to 'see' other wavelengths allows us to construct colour-coded images from radio wave, ultraviolet, X-ray and gamma ray radiation data (a technique referred to as applying 'false colours' [6-27]). Having this ability is like being given access to the 'Illustrated Handbook of the Universe' with all sorts of useful colour pictures and supporting text instead of the previous blurred monochrome images that defy proper interpretation. For me, the visual evidence now available provides yet more clear evidence that the EU model for star birth and their subsequent behaviour, is confirmed.

Now what seems to be needed is for those of us in the category of 'interested non-professionals' who care to consider the evidence using fundamental scientific principles, to grow in number and to pressure the astro-science establishment to show a more inclusive attitude towards electrical and plasma theories.

Is this a pipe-dream? I do not know, but what I can say is that I will personally do what I can in my own small way to help bring this about. It is an amusing thought to consider that the legendary Sherlock Holmes would have seen right through the weakness of the almighty gravity story and that he would have constructed an assessment of the EU model as the party actually responsible for the wonderful universe we have! The idea that gravity-initiated implosions and explosions of stars being able to expel mechanical shock waves powerfully enough to make dust and gas glow to emit the most harmful radiation in the universe is bunkum, hogwash, claptrap, and codswallop … in my opinion!

**Pulsars, Neutron Stars and Magnetars:** These supposed different 'types of star' and other related star types are treated by traditional astro-science as mysterious, but at the same time, understandable enough to describe to the public in confident terms. This has meant that the information given to us about them has left the impression that they really are all stars of different types. It will become clear, however, that they are all 'effects' rather than objects distinct from other stars, in fact, EU theory suggests that all of them may arise from electrical activity in the magnetospheres of otherwise ordinary stars.

The main discriminating factors for the powerful events we detect in space should now be apparent, these being the differing types and amounts of ionised matter, energy and time. Time, of course, plays a part in every event, but in the case of the effects being discussed here, it is appropriate to highlight its role in particular. Remember that events in the universe happen at a pace that is apparently so slow we do not think to consider its effect. We have been observing deep space and the events out there for nowhere near one hundred years, so again our arrogance and lack of appreciation of the actual scale of things blinds us to the fact that our judgement of these events, which we effectively see in still picture form, is in no way good enough to inform us fully of what the movie we are watching is really about.

The observed behaviour of so-called **Pulsars** has given them the name 'lighthouses of the universe'. This has been a handy analogy to use, but is misleading in the impression it conjures up of a rotating beam of light. Pulsars in standard theory are incredibly dense, narrow radiation beam emitting, rotating bodies formed through the dying star notion of a supernova event. They are labelled 'lighthouses' because they are observed producing pulses of EM radiation (X-rays) on a very regular basis. There is actually no direct evidence of actual rotation that would bring about a sweeping lighthouse beam effect; it is only because they pulse that physical rotation has been assumed. This is old and narrow thinking again! The radiation they emit can be identified and measured, but this is pretty much all that is known about them; everything else has been guesswork, and as with so many other things, nobody has seen one as an actual object in the eyepiece of a telescope! The idea of a beam emanating from a rotating body is immediately suspect when in the context of our plasma universe we consider a simple electrical explanation for the behaviour we see. However, before we do this, we will have a further brief look at the standard story around pulsars.

I think it is reasonable to ask, how, out of the thousands of pulsars detected so far, all of their assumed narrow beams of radiation are able to find Earth in their sweeping path. This suggests that pulsars must be extremely common in space. I do admit this would be possible, especially since their claimed narrow beams would spread out to become broad cones over distance, but there is no firm evidence to believe any of this is true. Moreover, astro-science seems to assume that all of the power emitted by pulsars can be detected and measured accurately. What then about the situation where the Earth may only find itself swept by the edge of one of these cone-like beams? How in that case could the true radiation levels measured be considered accurate enough data on which to make absolute judgements about the object from which they came? Mere 'convenient analogy', the mark of a doomed approach, seems to take the lead with astro-scientists today. The only exception to this is when ever more complicated mathematics indicates something might just be real, so if it suits their purpose, they typically go ahead and make it so! The mathematical invention of a rotating body with a narrow sweeping beam of powerful radiation is rejected in favour of what common sense EU theory has to say. In this respect, I want to explain a simple electrical circuit that provides the same visual effect that a pulsar does as it demonstrates the electrical principal that probably lies behind their observed behaviour.

This is a simple circuit in which the neon bulb on the right flashes repeatedly (pulses), just as a pulsar does.

Circuit of a simple relaxation oscillator © author

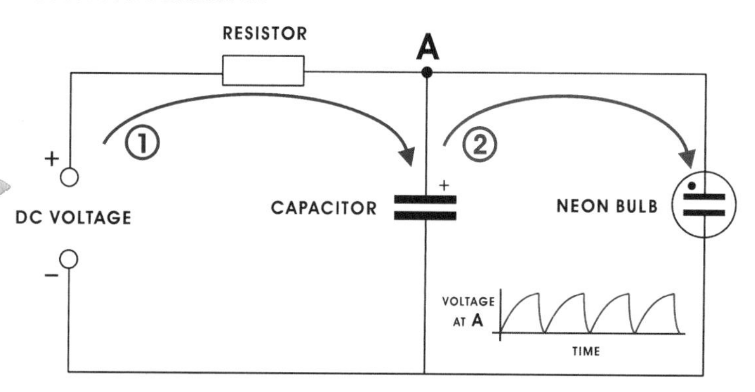


(Refer to the small inset graph.) (1) Current flow from the DC source is limited by the resistor so time is needed to build up charge in the capacitor. (2) When the charge in the capacitor reaches the 'striking or breakdown voltage' of the neon gas inside the bulb, a shorter time is taken for current to flow (discharge) from the capacitor through the gas to briefly light it up; this also lets the current make its way back to the negative side of the supply. The key here is the 'striking or breakdown voltage'. This can be monitored at point '**A**' where we see the waveform of voltage reaching an appropriately high value that will switch the neon gas inside the bulb from its plasma 'dark mode' to 'glow mode'. When this happens we see the brief period of glow mode as a flash of light before the gas reverts back to its dark mode. After a discharge takes place in the neon gas, the circuit will go on to do the same thing over and over again until such time as the DC supply is removed. Might this flashing action be mimicking the behaviour of a pulsar? Yes it does, and this is what is actually happening out in space, except on a scale that is hard to imagine. What we see here in this simple circuit is basic electrical engineering in practice. It is the same level of theory that is taught to first year electrical engineering students in colleges.

Here is what this experiment infers is happening in space. A voltage difference will exist between layers of plasma in the 'atmosphere' of a star, forming a Double Layer between them. On occasion, this voltage will build to a point where the double layer "explodes." (The complex signal expected from such an event has been successfully modelled by electrical engineers but their work is ignored in favour of the more outrageous hypothetical 'neutron star' model.) As the event proceeds, a pulse of visible light and X-ray radiation is released in all directions. One could imagine this as a cosmic lightning flash. Because the energy supply behind this is practically without limit, the charge will rapidly build up again across the DL to repeat the breakdown part of the cycle once more; and on and on it will go. We know from experiments and experience that DLs quickly self-repair after a breakdown occurs. It is this same action that takes place around Earth when in our upper atmosphere, electrical breakdowns occur that allow charges to build-up in our lower cloud layers which in turn discharge to Earth as bolts of lightning. (Lightning is not well explained by today's meteorology, its real cause goes back to the very interesting subject of 'space weather', but that is for another time!) So, to summarise, pulsar activity comes about as a result of electric breakdown action due to the build-up of excessive EM energy in a star's magnetosphere. A pulsar's apparent physical location can be thought of as the position of the 'Neon Bulb' component in a stellar scale relaxation oscillator circuit [6-28].

We are currently safe and cosy here within the Earth's biosphere where the powerful activities that go on in our dynamic electric universe are kept well away, and so they are essentially unknown to us. Those powerful events would be far more obvious if the DL barrier around the Earth was suddenly to be exposed to greater external electrical stresses and change in a way that made it less effective for our protection. If that happened, we would observe many new and probably terrifying events in our skies that would bring about powerful discharges to ground that would carve, bore and split it in ways that would be hard for us to conceive. This, of course, is unlikely to happen in our lifetime. Our experience of spectacular things such as pulsars is actually a short period in their total existence, and we have no idea how their behaviour may change over the great swathes of time that lie ahead. However, one thing is for sure, their behaviour and that of other phenomena like them is destined to change as the energy input that defines their current action changes.

**Neutron Stars**: As was said before, the description of these implies they consist of the heaviest (densest) form of matter in the universe; solidly packed neutrons with no space between them. This is the fabled 'Neutronium' of which a spoonful would weigh a billion tons here on the Earth. Hmm, perhaps, and only if it were possible to produce that stuff in the first place! The reason behind neutron stars being considered real objects arose from the need for an exceptionally powerful gravitational source that theoretical mathematics had said was required to explain powerful radiation emissions and gravitational effects that were observed coming from deep space.

The result of this mathematical assumption was the presumption of an actual body that possessed super gravity and also that this body could spin very rapidly while physically remaining together to emit copious amounts of deadly radiation as particles were drawn to its surface. This is what the standard story basically describes neutron stars as being. The origin they are assumed to spring from begins as the compressed core remnant of a very large star when it dies; a supernova event. (Please remember, this is according to the standard story and we will stay with this for just a little longer.) The collapse of very large stars is said to initiate the process of extreme atomic compression which in this case breaks atoms apart into their constituent sub-atomic components of neutrons, protons and electrons. At this level of structure breakdown the protons and electrons are forced to recombine to become 'new' neutrons. These, together with the original neutrons of the star then form a solid mass of neutrons which is claimed to have absolutely no space between those then cosy particles. What results is a most perfect sphere of neutron material that is calculated to be between 10 to 20 kilometres in diameter.

Most neutron stars are also claimed to have a high rate of spin associated with them. Here you can think of them being like pulsars emitting high frequency EM radiation but without the detectable beaming action. This leads to an interesting claim by astro-science that beggars belief as it highlights the extent to which they allow their imaginations to go. Despite the laboratory proven fact that neutrons cannot remain together or even remain stable individually for any more than 15 minutes, it is claimed that many of these kilometres in diameter bodies can actually spin at a rate of many tens of thousands of revolutions per minute and stay together. To highlight the basis of this extreme claim, if we take the diameter for the (typical) neutron star J1739-285 as being, say, 15 kilometres, can you imagine this incredibly huge ball spinning as is claimed at 51,720rpm? Well, that is what is being asked of us. This would mean that the outer surface of that sphere would rotate at 47.1 x 862 (i.e. the circumference of the star in kilometres multiplied by number of revolutions per second) which provides us with a figure of 40,600 kilometres per second or 146 million kilometres per hour! I leave you to consider this.

The bottom line is that neutron stars have never been 'observed' as anything other than high frequency EM radiation sources and apparent gravitational anomalies in space. They are only theorised to exist and have never been observed in physical form. In EU terms they are an 'effect' rather than an object; one that is linked with supernovae events that have at their heart typical Z-pinch events. These in turn produce stellar scale EM forces that act on and within dense plasma to bring about the effects now mistakenly attributed to gravity [6-29].

**Magnetars:** These again are said to be tremendously dense objects which, this time, are detected through the outrageously strong magnetic fields attributed to them and the effect these fields have on matter within their reach. Magnetars have only been hailed as distinct and 'real objects' in the past few decades, yet astro-science was quick to say they can be satisfactorily understood. However, the effects apparently observed that have encouraged their invention are again the seeds from which fairy-tale objects are grown. They are, in fact, understandable through the same creation methodology as provided to explain pulsars and neutron stars.

This time, however, the types of matter involved and the extreme nature of the Birkeland currents present have resulted in a concentration of EM force that produces magnetic fields of stupendous power and range [6-30] [6-31]. By addressing pulsars, neutron stars and magnetars in this way, I am conscious of possibly venturing close to promoting them as unique physical objects, which would be wrong to do. In reality, none of them exist as the objects that standard astro-science would have us believe. All of them instead result from powerful electrical actions that take place within regions of concentrated, charge separated ionised matter.

**Black Holes:** The story around gravity 'monsters' does not stop with what has been said to this point for there remains the supposed daddy of them all - Black Holes. Following on from what was said in chapter three, the observed phenomena of Black Holes seems to be an extension of what has already been described as the actual make-up and operation of pulsars, neutron stars and magnetars. It looks like all of these arise from the effects of powerful current flow within and between concentrations of plasma. The mistake being made, in my opinion, is that separate identities are being assigned to what are actually electrical actions that occur at every scale throughout the universe. If one looks closely at the subject of Black Holes, then in addition to being asked to stretch one's imagination in terms of what they are capable of doing, we are also in danger of being led to believe in the physical reality of 'nothingness' and 'infinity'. While admitting, of course, there is yet much to learn, this is still a stretch for anyone trying to keep their feet on the ground while attempting to come to a common sense view of our universe. Should we just let the theoretical astro-scientists and mathematicians get on with it and accept what they tell us is true? I say no, and I say we should pay no heed to the subject of Black Holes because they do not exist; even Einstein did not support the idea of them!

As long as today's astro-science community ignores the existence of electric currents at every scale in space, their examination and assessment processes will remain inadequate and misleading, not just for that community but for everyone. Their toolbox is incomplete and they are unaware of that fact, so all the problems they come across are 'fixed and explained' using the limited gravity tools they have. And unfortunately, they seem doomed to continue producing pointless conclusions from their predominantly public funded work. It is as if we were watching the comedy trio Larry, Mo and Curly (The Three Stooges) as pretend astro-scientists baking a simple cake - what should we expect them to produce if they do not include the flour they deny is needed whilst being in possession of clear evidence that says that it is?

The single limited snapshot we currently have of the universe is nothing in the grand scale of things. Despite being aware of this, we still seem happy and confident in making judgements and predictions about absolutely everything while showing no appreciation for the fact that a wider range of variously influenced events could have taken place during the time that has passed till now. We are undoubtedly a very capable lot, and I fully understand the inherent need we humans have to drive forward in everything, but is it not arrogant of us to be so certain of what we know at this point in time?

**Redshift:** Astro-science's interpretation of 'Doppler redshift' is the mainstay of the Big Bang (BB) and expanding universe stories. It turns out that through the narrow assumption that current redshift theory is correct, its evolved interpretation is said to prove that everything in the universe is moving away from everything else and that this has been going on for a calculated 13.7 billion years. Ostensibly, this implies that everything must therefore have originated from a single point at that far distant point in time - the moment of creation!

The highly respected astronomer Halton Arp, the man who made it part of his life's work to collect and analyse images of unusual deep space galaxies, has provided evidence over decades that shows the use of redshift, as a factor in the measurement of astronomical distances, has likely been a wrong thing to rely on. Credible evidence now exists that shows redshift, as defined and used by standard astro-science, should not be used to support the assumption that distances can be calculated reliably or that the universe is expanding or indeed, as proof that a BB type event actually happened. These are significant implications, and they further imply that setting the age of the universe at 13.7 billion years has also been a mistake [6-32A] [6-32B].

Arp's evidence and data analysis shows that redshift, instead of being 'one thing', actually has two components and that these should be separated for either of them to be valid and useful. Here we have the apparent motion related 'recessional' component that represents the stretching of wavelength that would happen with light coming towards us from deep space objects if they were moving away from us. Then there is the much more significant 'intrinsic' component, which is apparently related to the object's age.

There are two things of great relevance to note here. The first is, that the aspect of an intrinsic component has been ignored by astro-science - they will not clearly recognise it exists or that it relates to the age of an object (particle physicists tell them it is not possible.) The second is that without attempting to separate the intrinsic value from the total redshift value to leave only the recessional figure for their distance estimation sums, they have stubbornly retained the intrinsic component in their calculations that supposedly 'prove' the recessional speed and distance from us of deep space objects. Further to this, it may well be that there is little recessional aspect to redshift at all and that the intrinsic value has the most to tell us about redshift's implications.

These things are very important indeed because results from traditional redshift calculations have been used to formulate and support major astro-science theories that have mistakenly been assumed as accurate and acceptable by almost everyone. Furthermore, these 'facts' form a basis of assurance for aspects of certain belief systems, education programs at all levels and major research programs and commercial investments. The upshot of this is that if redshift values are not representative of recessional speed and distance, then the universe is not expanding and the idea of a Big Bang event is wrong [6-33].

By way of providing evidence to back this up, consider the next image of the low redshift galaxy NGC4319 at a calculated distance of 80 million ly (light years) and the high redshift quasar Markarian-205 at 1 billion ly. (Note, 80 million versus 1 billion has a factor of 12.5 times difference.)

Visually, it appears to us that both these objects are physically connected by some form of matter bridge, even though standard redshift calculation of the distance each of these objects is away from us tells a different story.

The point being made here is clear; if redshift does not relate to distance, as the work of Arp suggests, then how can standard astro-science say anything at all about the distance away from us of these two objects, or indeed the distance from us of any other object in space?

In this example, the two objects seem to have an obvious connection so should be assumed to be located at the same distance from us [6-34].

Credit and copyright: John Smith, Hidden Loft Observatory, Don Scott

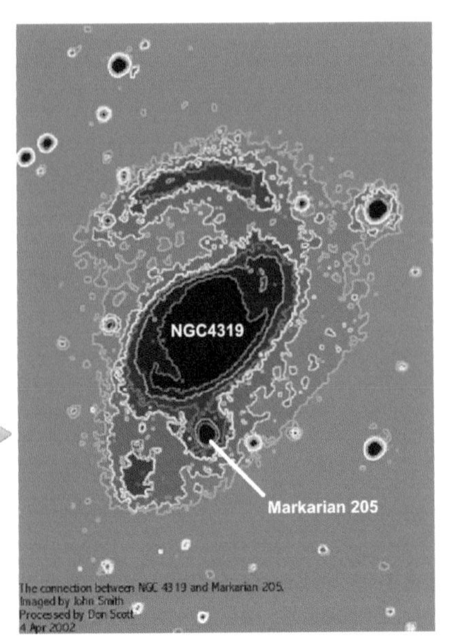


Another example is a highly redshifted quasar that appears to be in front of the low redshift galaxy NGC 7319. The problem is that the calculated redshift values for both the quasar and the galaxy tell us that the quasar is supposed to be 90 times further away than the galaxy is; we should not even see this quasar [6-35].

Credit: NASA/Hubble Space Telescope

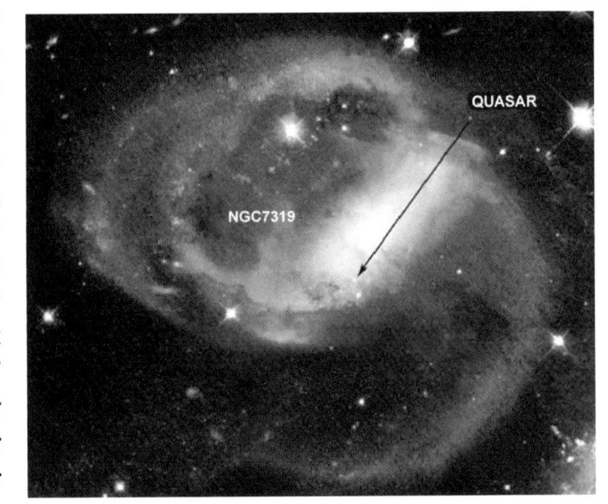


There exists a substantial amount of further evidence that shows physical connections do exist between active galaxies and quasars, which according to standard astro-science calculations should be whopping great distances apart. Would it not be fair to expect that some serious attention be paid to this? Sadly, it is not, for to do so in public would be to risk diverting the course already set by the currently comfortable helmsmen of the astro-science ship. It is my opinion that these intelligent people, many of whom will have suspicions and even knowledge of what is really going on, do not want to upset their 'professional' work with its often guaranteed research funding and veneer of peer respect. After having been left without proper accountability for so long, I think every aspect of their fragile worlds would collapse, and they are aware of this. This does not help the honourable pursuit of good science.

To expand a little on the 'intrinsic' component of redshift, I will follow on from the description of **quasars** begun in chapter three where I mentioned redshift as well. The EU model's explanation for quasars as distinct bodies is that they are yet another aspect of powerful electrical energy interacting with plasma, this time on a galactic scale. I ask you here to accept that Birkeland currents continue to flow into the cores of galaxies to maintain their operation and structure, and through a subordinate network internal to the galaxies themselves, the stars within them.

A pair of quasars being emitted from an AGN © author

The Birkeland currents that focus at the core of a galaxy, those which originally produced it and go on to maintain its central plasmoid, will be subject over time to fluctuations that are likely to cause the build-up of excess ionised matter and EM energy within the galaxy's plasmoid core. The result, just as with stars, will be for those over-stressed cores to eject some of that accumulated plasma. This it would seem is how quasars are born [6-36]. When a galaxy ejects material it does so in a single action that involves ionised material being expelled as spinning blobs from both its 'top and bottom sides'. The material expelled then spins away in opposite directions, probably along the axis of the two incoming Birkeland currents. This is why quasars are often observed as if they are organised in strings that lead away from the centres of galaxies that have a powerful active nucleus or AGN - this is the reference commonly used to represent a galaxy with an Active Galaxy Nucleus.

How 'z' is associated with quasar development © author

Newborn quasars have very high redshift 'z' values while their parent galaxies have relatively low values. This is a further clear reason to disbelieve the claim that redshift represents distance. In terms of quasars, the 'intrinsic' component of redshift is instead accepted as a measure of their youthfulness. Redshift is also found to have associated with it, a spread of distinct z values that have no other significant values in-between. When a young quasar is ejected from a galaxy, it initially has a high intrinsic redshift that seems to decrease through this sequence of values as the quasar gets older. A quasar also has an initial high 'ejection speed', but this reduces as it starts to take on the shape and features of a miniature galaxy and as it slowly assumes an electrical form of balance in a location near to its parent galaxy. This is how many of the galaxies we see appear to have small companion galaxies. The concept of intrinsic redshift value is also involved in a further interesting observation [6-37].

With reference to the graphic here, no matter in what direction we look in the sky to observe quasars that often accompany galaxies, they all seem to share the same range of distinct z values. If as is claimed, the total redshift of a quasar were to truly represent distance from us, then by this measurement, all quasars would be located within well defined shells of increasing set distance from us. This places Earth at the centre of everything, which of course is nonsense and another reason to see that the traditional interpretation of redshift is misguided.

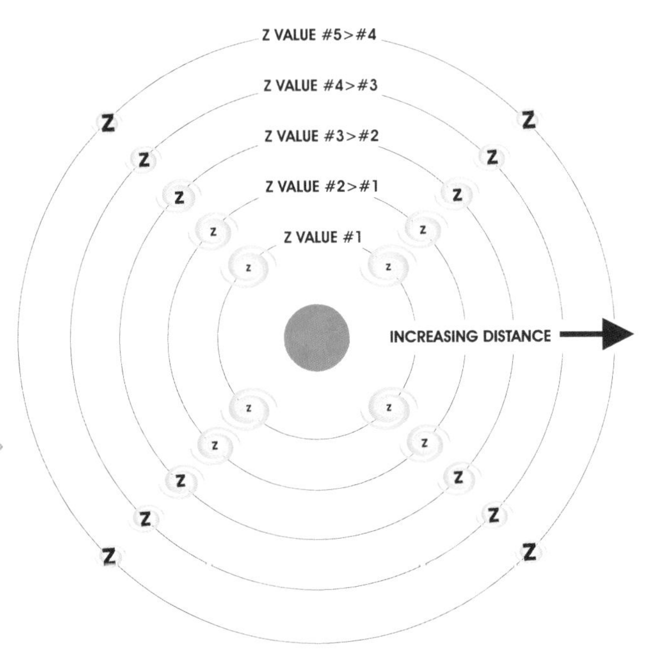


The nonsense of quasar 'shells' of ever-increasing distance from Earth © author

I mentioned some personal detail about Dr. Halton Arp in chapter five. Even though his own work and the gathered evidence from the work of many others is so substantial, he and they remain sidelined. [ Dr. Arp's website: www.haltonarp.com ]

## More on the electrical properties of charged bodies

We will be considering shortly the electrical properties of comets, asteroids and meteoroids, but first, I need to take you back to electrical theory - nothing too complicated, and I will attempt to explain things clearly. I will begin with an imperfect but adequate analogy to describe how the electrical component known as a 'capacitor' works in a direct current (DC) circuit. Think of this capacitor as a small bucket with a hole in its base. We are able to put water into the bucket, but only through this hole, and the hole is the only way that water can leave. It is obvious that depending on the pressure of the water flowing through the hole and into the bucket that a certain amount of time will be needed to fill the bucket, and in reverse, depending on the amount of water in the bucket that defines the pressure, it will take a certain amount of time for that water to run back out. If the bucket were big and we had the same size of hole as for the small bucket and also the same supply pressure of water, then the filling and emptying times of that larger bucket would be much longer. However, the total amount of water eventually stored in it (which we will come to think of as the amount of energy) would be greater. Let us accept this idea as an adequate starting point.

We now replace the idea of a bucket with our electrical component, the capacitor. You will see here that the role of the capacitor is just the same. Consider the flow of the water as being equivalent to the flow of electric current. On the way in to the capacitor, the current is called a charge current, and on the way out it is called a discharge current. Just as the bucket is a container for water, the capacitor is a storage location for electric charge. Here, the pressure (voltage) driving the current flow will determine the time taken for the capacitor to take on its full charge. Small capacitors will charge up quickly, and large ones will take much longer. When we consider the discharge of capacitors, this process will take an amount of time determined by resistance to current flow in the circuit of which the capacitor is part. If resistance is high, then a lot of time will be taken for it to discharge. If resistance is low, especially in the case of a direct short circuit, the discharge time will be extremely short - and depending on the amount of energy that has been stored, a short circuit may produce a very powerful electric discharge event indeed! (Here you can associate this with the image of turning the bucket of water upside down and letting all of its contents out at once.) It must be noted, however, that some form of complete circuit is always required for charge and discharge cycles such as these to take place. In real life, the configuration of this 'complete circuit' may not be obvious at first, but a complete circuit there must be. The charge for the capacitor in our simple example comes from a standard battery, which is a steady DC source (any external steady current source would do the same job.) Keep in mind as we go through this example that we will shortly be considering situations where the action of the circuit described is the same as that which goes on between regions of differentially charged plasma, and between charged bodies like planets, moons and comets.

In stage (1) of this simple circuit the capacitor will charge from the battery when the switch is in position '**A**'. In stage (2) when the switch is changed to position '**B**' the stored charge in the capacitor will flow through the circuit resistance or perhaps through a short circuit (which is a negligible resistance) back to the other side of the capacitor The charge circuit is shown in **Blue** and the discharge circuit is shown in **Red**.

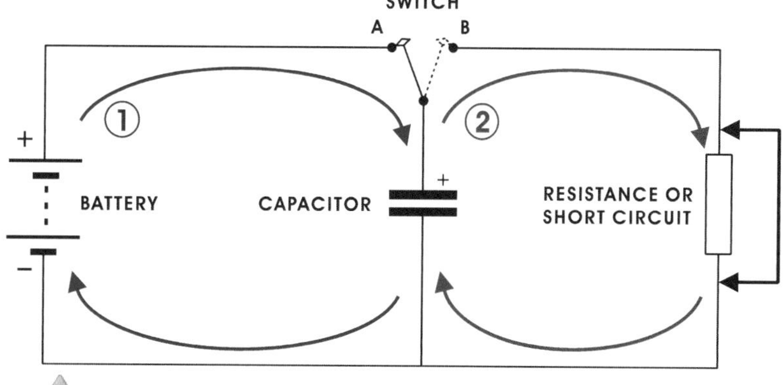


Circuit showing the charge and discharge cycles of a capacitor. @author

In real life, a capacitor has two metal plates that take on separate negative and positive charges. This is normally because of voltages being unequal within sections of an electric circuit between which the capacitor is connected. The plates of the capacitor are separated by a physical insulator, or perhaps just an air gap which serves this same purpose and across which an electrostatic charge can build up. Let's forget the capacitor as a physical component and consider only its ability to store an electric charge. Think of your own body as one of the metal plates of a capacitor that can take on and store a charge. A rather intimate implication of this becomes readily apparent when you remove an item of nylon clothing, where the action of doing this unknowingly bestows on you a significant electric charge. You then proceed to touch a door handle and get an electric shock. You can think of this second surface, the door handle, as serving the same purpose as the other metal plate, with the air between you and the handle acting as the original insulator. By touching the handle with your hand, you have removed the involvement of the insulator (the surrounding air) and have allowed the charge on your body to flow through the door handle to earth. In a case such as this, your body will have had a much greater negative charge than the door handle, which is the same as saying that the door handle has a much greater positive charge than your body. The fundamental point is that any two bodies or regions of previously separated charge will behave in the same manner to some degree if they each have different charge levels and then are brought close to or in actual contact with each another. For our purpose here, 'capacitor action' allows energy storage because it keeps charges separated until something external changes the physical situation or if the charge difference between the two bodies or regions gets too great and an electrical breakdown across the separating insulator occurs. By scaling this concept up to accommodate all solid bodies and configurations of charged plasma at every scale in our Electric Universe, capacitor action is fundamental to our understanding of the theories involved [6-38].

**Comets** are solid bodies that in our experience are very much smaller than planets and moons. The standard description says they are dirty, cold snowballs left over from the formation of the solar system and that their long elliptical orbits which ensure they are rare visitors to the inner solar system are due to them being disturbed in the distant past from either the Kuiper Belt or the theorised Oort Cloud collection of comets. The force that has disturbed them is typically attributed to the gravitational effect of larger passing bodies or a chance collision with one of their companions. This is what standard thinking on comets would have us believe. Contrary to this, we now have evidence that clearly points to typical comets being electrically charged rocks left over from catastrophic collisions and/or destructive inter-body (inter-planetary or moon) electrical discharge events in the past. Some of these events hail from the relatively recent history of our solar system, and some from its much more ancient past when our planets and moons interacted violently with each other. The orbits of what are called short-term comets take them from the Kuiper Belt that lies beyond Neptune to a close passage with the Sun. The long-term variety also pass close to the Sun but are said to originate much further away in a supposed shell of icy bodies that encompasses the solar system and which has been named the Oort Cloud. This is imagined as a solar system encompassing collection of icy bodies that is located much further away than the orbit of Pluto.

This long-term variety of comet travels continuously between the far away negatively charged environment of the outer heliosphere into the Sun's very positively charged central region. As a negatively charged comet comes close to the Sun, it isolates itself electrically more and more by developing a DL barrier around it to separate its charged body from the Sun's increasingly positive environment. A comet body's reaction to this often rapid change in electrical stress is for its own plasmasphere to start to glow brightly as electrons flow from the comet's surface to reinforce the double layer between its plasmasphere and the positive heliosphere outside. When a comet's plasmasphere is affected in this way it glows and is called the comet's **coma**. Remember here that it is the size and composition of a comet's rocky central body, its nucleus, that will define the charge limit it has and the rate at which charge is taken on or removed (dissipated). This provides a basis on which to explain the observation that some small comets can swiftly become very bright indeed as they encounter regions of highly ionised plasma. This flaring action can easily occur very far away from the Sun [6-39].

This explanation clearly holds up because what we observe and deduce about comets in the solar system can be reproduced in the electrical laboratory. In contrast to this, the 'dirty snowball' model is now viewed as bad guesswork that has never satisfactorily answered even the simplest of questions about observed comet behaviour. Questions like … Why do recent close-up pictures clearly show a rocky and sharply defined surface, where instead, a melting ice surface would definitely not appear this way? Why are comets the blackest of black objects in the solar system, appearing as if they have been burned? Why do we actually observe bright spots (electrical arcing?) on elevated and sharply defined features on the surface of comets? And, why in 2005 did NASA get their '*comets are made of ice*' predictions completely wrong concerning the Deep Impact mission sent to comet Tempel 1? In particular regard to this, quietly and professionally, the physicist Wallace (Wal) Thornhill of Electric Universe renown got it absolutely right with his own set of predictions for electrical behaviour that would be observed on and around that same comet. NASA had launched this mission to fire an 820 lb copper projectile at comet Tempel 1 so that the impact and the resulting ejected debris could be studied through radiometric and visual analysis. Wal's main predictions turned out to be correct [6-40], but went unacknowledged by NASA scientists and engineers because practically all of their own were wrong. Among other things, he predicted that a flash would occur prior to impact; the impact itself would be much more energetic than what a purely mechanical collision could bring about, and 'jets of ejected matter' would be observed moving around, probably to accommodate the redistribution of electric charge on the comet.

The evidence that comets are actually negatively charged rocky bodies that endure massive electrical discharge events as they interact with the Sun's electric circuit is mounting fast. Due now to better data, the quantity and quality of questions about comets is becoming harder for the astro-science establishment to ignore. Given that we can now study comets very closely and reveal their undoubted electric behaviour, then in terms of astro-science's continued rejection of the notion that electric currents actually do flow in space, it is possible that comets might turn out to be the straw that breaks the camel's back, so to speak.

The Kuiper Belt and the Oort Cloud - what are they really? The Kuiper Belt lies beyond the furthest gas planet Neptune and is said to consist of many icy and rocky objects of various sizes up to small planet dimensions. Pluto, being in that same region, is considered a Kuiper Belt Object (a KBO) that still has a debate going on around it as to whether or not it is actually a planet. Well, whatever classification it ends up with, it sure is a big object. A number of other KBOs have been observed to which we have assigned names such as Sedna, Orcus, Quaoar, Ixion and Varuna. The evidence is that many KBOs are very substantial objects, but we know little else about them. Despite everything being very cold indeed in that region, no objects out there or anywhere else have been truly determined as floating chunks of ice; the existence and dominance in number of icy bodies in these locations is all just guesswork. The Kuiper Belt is the home of short-term comets such as Halley and Swift-Tuttle, both of which regularly visit the inner solar system every few decades. This type of comet is not the main focus of the information to come because short-term comet orbits do not see them travel too far away from the Sun. This means that the electric display they put on, for which I will explain the reasons in terms of all comets, is often not as dramatic as what happens with the long-term variety [6-41].

The Oort Cloud is said to be where long-term comets come from. The physical form of this 'cloud' is described as a thinly populated globe-like shell that encompasses the whole solar system at a distance of 7 to 25 trillion miles away from the Sun. This is a enormous structure, virtually impossible for us to imagine. Comets originating from there are said to take between a few hundred to millions of years to complete a round trip past the Sun. This includes comets such as Hyakutake and Hale-Bopp.

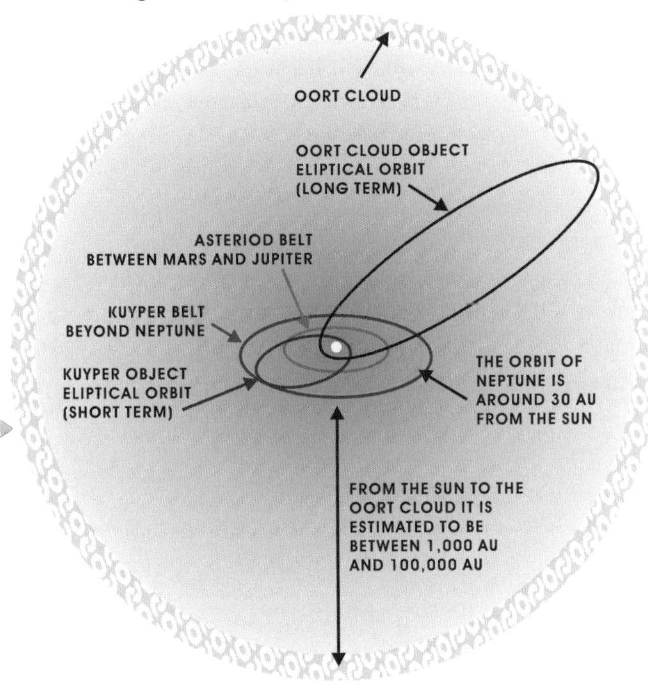


The orbit paths and limits of short and long term comets © author

The suggestion that the Oort Cloud is like a globe comes about only because long-term comets are seen as coming towards Earth from many different directions. Therefore, astro-science has judged them as originating from an all-encompassing shell-like structure that exists in a very cold region of space indeed. We must note, however, that no Oort Cloud objects, tiny though they are supposed to be, have ever been detected in any way whatsoever; they have been assumed to exist, so we are again faced with being asked just to believe what we are being told [6-42].

In both cases where we see long and short-term comets coming into the inner solar system, the regularity of this seems far greater than their estimated numbers or their possibility of being disturbed by passing stars or collisions with other bodies would suggest is possible. However, there does seem to be some form of influence out there that so far has been left out of the explanation for how long-term comets are sent sunward. Perhaps the previously suggested idea of a brown dwarf companion to our Sun could be the answer - who knows?!

A rapid change in a comet's level of charge can be such a catastrophic experience for that body that very violent things can occur both internally within its nucleus and externally on its surface. This is especially the case if a comet finds itself close to another body's plasmasphere that has a very different charge level from its own. If this happens, comets can break apart due to fracturing by internal electrical activity (think of this as internal lightning through rock.) Apparently, this is what happened to comet Shoemaker-Levy 9 in 1994 when it got too close to Jupiter and fractured into 21 separate pieces [6-43].

Credit: H.A. Weaver, T. E. Smith (STSI) and NASA

Other powerful events can occur as comets travel within the heliosphere through regions of dense solar plasma that have been ejected in bursts from the Sun when solar flares or CMEs (Coronal Mass Ejections) take place. In addition to the ambient drift of the solar wind away from the Sun, the plasma ejected in concentrations by solar flares travels at hundreds of kilometres per second past the inner planets and continues to accelerate away. The patchy regions of charge in this environment are able to have a powerful effect on in-coming and out-going comets, even though they may still be very far away from the Sun.

When a comet approaches the Sun its plasmasphere becomes visibly active and glows as a coma because it switches from the dark mode of plasma to glow mode. Depending on the levels of charge involved, the coma can become very large indeed. Comet Holmes displayed this behaviour in 2007 by showing off its 2 million kilometre diameter coma, which amazingly was larger than the diameter of the Sun [6-44].

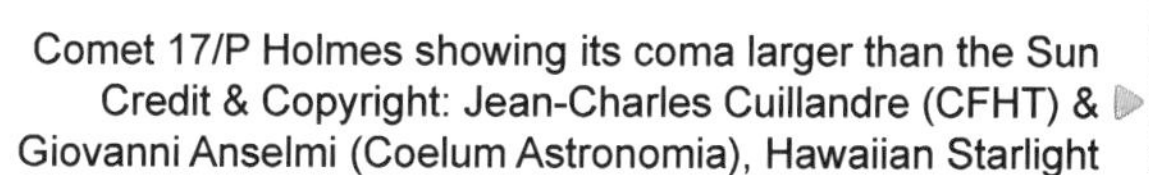

Comet 17/P Holmes showing its coma larger than the Sun
Credit & Copyright: Jean-Charles Cuillandre (CFHT) &
Giovanni Anselmi (Coelum Astronomia), Hawaiian Starlight

As mentioned previously, comet Holmes' core (its rocky nucleus) was estimated only to be 3.4 kilometres in size. (The white dot you see in the middle of this picture is the star Mirfak shining through and not comet Holmes' nucleus.) Think about the numbers involved here, and I will say it again, what utter nonsense for NASA and astro-scientists to persist in attributing observations such as this to dust and gas being thrown off the surface of a tiny fragment of ice – really!

To divert for a moment. Comets have always brought wonder, amazement and sometimes fear, especially for people in ancient times who only had limited experience to draw on when they interpreted them as harbingers of dreadful events when they traversed the heavens. This situation appears in sharp contrast with what we are able to say about comets today due to all the fine equipment we have available. It is unfortunate, however, that astro-science has produced a blinkered and confused description of comets from the vast resources it has, and this leads me to consider in a new light the simple and honest motivation of people in ancient times who told stories of comet visitations. Perhaps some of those stories involving comets that described battles in the skies between their gods and heroes and great destruction being wrought on Earth were their only way of relating what they thought was really going on. The ancient view of comets is a fascinating and revealing area to look into.

There are times when we jump to a conclusion against our better judgement and without restraint let everyone know what our thoughts are, then immediately regret we did. For whatever reason, some of us then stick to the original line in the naïve hope that the storm can be ridden out, unscathed, and that our bad judgement will not be exposed. Well, I liken this to the situation where we have the terminally-tunnel-visioned-tenacity of astro-scientists who hold on to the dirty snowball theory of comets. Any good evidence that strongly supports their model is absent, totally. In the past, the only evidence they put forward worth listening to is that the 'chemical signature' of water ($H_2O$) has been found in comet tails. There is now other strong evidence that this discovery has been a mistake in their interpretation of what is being found. To determine this they use a process that examines the properties of light that comes from comet tails. This is where they seem to have mistakenly interpreted the detection of OH [6-45] as being the $H_2O$ they were actually hoping to find. (OH is one oxygen atom plus one hydrogen atom - this is not water; it is a negatively charged molecule known as a *hydroxyl radical* that has an extra electron.) The process used to facilitate this interpretation ignores completely, any possibility of electrical forces being involved. This is especially frustrating as it is known that OH can be produced in the laboratory through experiments that mimic the positive solar wind's interaction with a rocky comet's negatively charged surface material. There, negative oxygen ions are produced and expelled by the electrical action that takes place on the comet's surface; these ions then combine with in-coming positive protons (hydrogen nuclei) from the solar wind to produce OH. Despite strong evidence that their analysis is wrong, they still seem to prefer the interpretation of OH as $H_2O$! I admit to taking everything on faith here, because I have no personal qualifications in chemistry. However, I do take seriously the detailed observations and analyses of notable scientists in their assessment of this and other bloomers that have been and are still being made.

Now, something about comet tails; there are usually two. The visually larger and often 'curved' one is said to consist of slower moving heavy dust and debris particles ejected from the surface of the nucleus as the result of ice sublimation. The other 'ion tail', the straight one, is said to be formed by the action of high-speed, high-energy protons in the solar wind striking gas particles expelled from a comet's surface with enough force to make them glow as they are ionised. The resulting brightly coloured display stretches out to a narrow tail behind. This is the standard story of comet tails.

Comet Hale-Bopp - Credit: NASA

In the Electric Universe there is a different view of comet tail production due to electrical action being responsible for them and not solar heating. The heavy dust and debris of the curved tail is indeed material removed from a comet's surface, but not via the action of sublimating ice. This surface machining process will be covered in detail very shortly but for now I will continue with the 'ion tail' description. In 1986, the Ulysses space probe detected evidence of the ion tail of comet Hyakutake at a distance of 360 million miles behind the comet itself - this is close to the distance between the Earth and Jupiter; a fact that gives us an important thought to chew over.

We should remember that space is a very good but not perfect vacuum and that under normal circumstances, electrically neutral gases like those claimed to be expelled from a chunk of sublimating ice would disperse very swiftly and not remain together as a structure that has form. Therefore, the question is, what could have held Hyakutake's ionised tail together as a long filament for it to be detectable at an enormous distance from its source? It certainly could never be gravity, magic or the fact that the molecules of gas were just being friendly, so we find ourselves back with the EM force to explain this. Here again we consider the constraining action of a surrounding magnetic field, this time the one formed by the Birkeland currents flowing within Hyakutake's ionised tail. It is that field that would naturally act to maintain the tail's narrow structure over the vast distance in question. If astro-science's idea of this tail consisting of neutral gas was true, then instead of forming a tail with structure, the gas would have rapidly and randomly dispersed into space after departing the nucleus of the comet [6-46]. Again, we find ourselves talking about the involvement of filamentary Birkeland currents. In the electrically dynamic environment of space, we can expect to find them everywhere, supplying energy to powerful events or being generated by the interaction of mobile charged bodies within their plasma environment.

There are many other visual observations and data records that contradict standard comet theory and which seem to support the logic and experimental results behind the alternative electric theory. Why in 1991 was comet Halley seen to 'expel' a cloud of glowing dust from its 15 kilometre nucleus, where this cloud then grew rapidly to be 300,000 kilometres in diameter, even though Halley was 14 times the distance from the Sun than the Earth is [6-47A] [6-47B]. What level of heat from the Sun could have been present to affect so significantly a piece of ice at that far distance? In this case, and as previously with comet Holmes, the explanation is electrical. The double layer surrounding the comet's negatively charged nucleus simply interacted with a significant concentration of positively charged plasma in the heliosphere (produced by a CME from the Sun perhaps) at that location.

Observations also indicate that some comets seem not to be very solid or that they are bodies with very little mass. Astro-scientists have taken this apparent lightness (fluffiness, as they call it) as evidence that comets are indeed snowball-like or that they may be internally 'honeycombed' in some way by empty tunnels and voids. Bear in mind that these athletically creative views are offered confidently despite the great amounts of evidence that contradict the standard model for comets. One wonders how this can be, especially where we have close-up images of comet nuclei showing rugged, rocky, cratered, gouged and pitted surfaces that bear no resemblance to ice in any of its forms in any way whatsoever. The apparent 'lightness' being detected can be accounted for *if* we are willing to accept that gravity itself is fundamentally electrical and just another aspect of the EM force. Quite a leap, I know, but there are good reasons to accept this as true.

The main piece of evidence here is that the weight (mass) of a negatively charged body located within the positively charged environment of space, would in the view of some really clever people, measure differently depending on the amount of negative charge it takes on. In terms of comets in space, this means that the highly negatively charged nuclei of comets travelling within the highly positively charged environment around our Sun, where the electrical difference between them would be very significant, would appear to us as if their weight was akin to that of a lump of polystyrene foam! I personally find the thought of electrically modifiable gravity fascinating, and if true, it could provide a route to consider explanations for other puzzles in astro-science and from archaeology and geology that have traditionally had a relationship with gravity; significant puzzles that have been around for a very long time. I underline, however, that this is my own thinking wandering off at a tangent [6-48A].

The craters and other sharply defined features that we see on the surfaces of comets can also be explained. Here once again we must keep in mind the scale of electric currents that are potentially available to flow between the plasma environment of space and the highly charged surfaces of comets. It may seem here tempting to imagine powerful Birkeland currents cutting into and lifting material away from comet surfaces, and this is fine to have as a mental image, for it is indeed what happens.

Consider the electricity available around us. Have you ever seen a powerful electric welding spark or a simple electric fuse wire blow in a sudden flash at home? If you have, then you will know that there certainly seems to be a lot of instantaneous energy involved. This is because significant levels of electric current will vaporise metal and send debris flying as sparks in all directions. It can also result in surrounding surfaces becoming blackened with fine soot and splattered with tiny metal spheres. This may be the only evidence left behind when an old fashioned fuse wire 'blows' in the home, for it is common for those wire safety devices to totally disintegrate.

Electric Arc Welding © www.electrical-picture.com

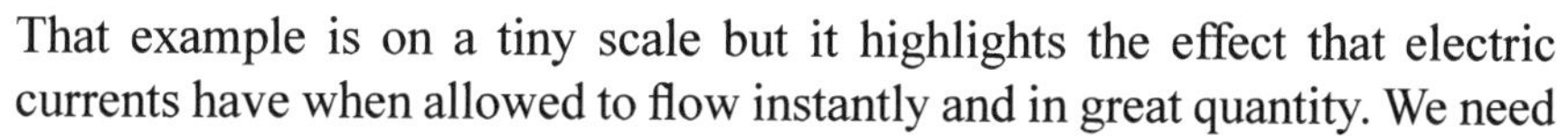

That example is on a tiny scale but it highlights the effect that electric currents have when allowed to flow instantly and in great quantity. We need also to keep in mind that this example would involve only a few amps of electric current at the level we use in the domestic setting and not at the scales of millions and possibly billions of amps of current that would be released in space. Now consider the potential effect of those exceptionally dense Birkeland currents on a planetary scale. The idea of these operating within our solar system could easily provoke a mental picture of great damage being done to the surfaces of our planets and moons. We would not need to expand this picture much further to also appreciate that material broken up by these currents and ionised could also be removed from the surfaces of comets and asteroids by such EM forces. Is it not then likely that the craters, channels, ridges and other geological features seen on comets and asteroids, might also be products of this process?

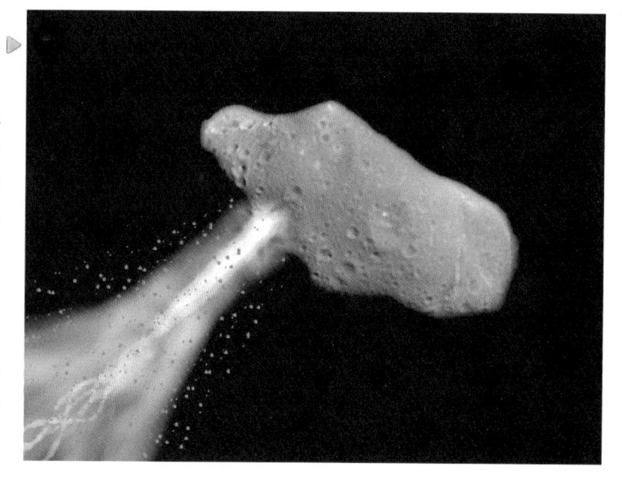

The concept of an asteroid or comet-like body being 'machined' by a Birkeland current © author

This is the type of event that indeed seems to happen when comets get near the Sun and become electrically over-stressed or when they get too close to other charged bodies that have vastly different charge levels. It surely appears through logic and observation to be the case that when voltage difference between areas of a comet's negative surface and its surrounding positive environment is great enough, then classic Birkeland currents will spin up from their surfaces to produce with powerful ease and also gentle erosion, the range of features we see.

Astro-scientists interpret these plasma discharges as jets of gas that escape from sublimating ice under a comet's surface. The facts tell us this is bunkum!

Comet Hartley 2 (EPOXI) with plasma outbursts from its surface
Credit: NASA JPL-Caltech UMD

As comets continually circuit the Sun, the electrical machining events described gouge out craters and gullies to leave ridges and mesas on their surfaces. The material removed is ejected into space as dust and larger lumps of debris, pieces of which are destined to become some of the meteors we see entering Earth's atmosphere. The amount of solid material removed would be impressive on our earthly scale. As their journey and this 'surface sculpting' process continues, the solid nuclei of comets are destined to eventually be machined away to dust. This is the same EDM process that our manufacturing industry uses today as a well understood technique for producing intricately machined metal parts through a controlled plasma discharge process [6-48B]. This process highlights proven technology that industry knows well and uses every day that can be directly associated with what we see taking place in space; the similarities are obvious. We must keep in mind that we have only been talking about comets here. However, as you will see, these same machining processes and discharge events can apply to all solid bodies in space, and it is especially interesting when we consider them in terms of the planets and moons of our own solar system. It seems that Mother Nature is a sculptress and her tools for that purpose are electrical.

Some of the clear images we now have of rocky comet surfaces show bright glows along the sharply defined rims of craters. This makes electrical sense because high-points on a comet's landscape are exactly where negative charges will collect and begin to 'arc' when the current density of the environment is powerful enough to force a slow but continuous discharge event - a bit like the process you may recall of smouldering paper when you blow on it and watch the glowing edges eat into the paper then turn to ash and fall away. In these better images we now receive, many of these 'glowing outlines' can be picked out for deeper analysis.

Thinking more about this type of discharge, have you ever seen those electric bulbs that are supposed to look like a flickering flame? These bulbs are filled with neon gas, and in normal operation, the voltage across the two electrodes inside will rise to a point where the ionised gas in dark mode will switch to its glow mode state as the gas rapidly conducts some amount of current, as previously described. By this action the plasma gives off a red-orange glow while appearing to jump around, so the appearance is that of a flickering flame.

This same 'jumping around' action of discharging plasma on a much larger scale is what occurs on the surface of comets when the discharges move from one raised sharp edge to another as the charge conditions around the comet itself change in response to the rotation of the comet nucleus, off-centre within its plasma sheath or coma.

Comet Tempel 1 with discharges concentrating on sharp edges
Credit: University of Maryland JPL-Caltech NASA

As this process continues the surface material is stripped, atom by atom, to form gouges, gashes, craters, ridges and mountains, either as single features or in groups that are often seen to have overall patterns associated with them. All of these features now easily observed on comets and other solid bodies are due to electrical discharge and machining events. There is more to come about this surface sculpting process when we discuss the surfaces of the planets and moons of our solar system.

Comets can actually be hot! Yes, I am afraid even NASA said this is the case in one of their 2002 articles [6-49]. In this article and on the basis of data received from the Deep Space 1 spacecraft mission, Dr. Laurence Soderblom of the U.S. Geological Survey's Flagstaff, Arizona said … '*The spectrum suggests that the surface is hot and dry. It is surprising that we saw no traces of water ice*'. Then in the same article there is this quote by Dr. Bonnie Buratti, JPL planetary scientist and co-author of that article ... '*Comet Borrelly is in the inner solar system right now, and it's hot, between 26 and 71 degrees Celsius (80 and 161 degrees Fahrenheit), so any water ice on the surface would change quickly to a gas.' … 'As the components evaporate, they leave behind a crust, like the crust left behind by dirty snow.*' So, first they say it is hot and dry and there is no ice, then the analysis gets support from the old snowball story. This seems at best to highlight a lack of clarity in their ability to describe what has actually been found, and at worst, a level of confusion and lack of clear communication ability [6-50].

Some comets have been found to give off X-rays, a form of high-energy radiation that cannot be explained at all by the snowball model. Again, EU theory can provide an answer. We have already established how electrons are stripped from the comet nucleus. When this happens they are accelerated by the strong electric field at the boundary of the comet's coma from where they go on to collide with solar wind protons. This process can cause radiation including X-rays to be generated. We have technology here on Earth that generates X-rays for scientific research and medical purposes, where one important method for this involves a device known as a 'Synchrotron', the radiation produced being referred to as synchrotron radiation [6-51].

**Asteroids** and comets are fundamentally the same things, however, the typical locations and behaviour of asteroids is different from that of comets.

The rubble strewn surface of asteroid Itokawa
Credit: ISAS and JAXA

Due to their locations and patterns of movement within the Sun's heliosphere, asteroids do not have the same chance as comets to experience significant changes in voltage with respect to the solar wind. They are large, sometimes very large pieces of rock that drift around the Sun in locations usually very far away from Earth [6-52]. Occasionally, however, we do hear of some that pass by us at relatively close distances. Asteroid '2009 DD45' at almost 150 feet across passed by at a distance of 45,000 miles in 2009, and asteroid '2004 FU162' at around 20 feet across passed us by at 4,000 miles distance in 2004.

No concrete evidence has been gathered about any large asteroids having struck the Earth in our recent recorded history, so it is understandable if we just accept, for now, that a large asteroid reaching the Earth's surface will be a rare event indeed. However, we should always remember that humans have only been around for an extremely short time, so what do we really know about asteroids? Therefore on the face of things, there has been no actual opportunity for significant asteroid impact events ever to be recorded in scientifically acceptable terms.

According to the EU model and just as with short-term comets, the place of origin of asteroids is out where violent electrical discharges during inter-planetary encounters have taken place in the past. In the long term, there remains potential for this to happen again, where undoubtedly, comets and asteroids would be products of those awesome events [6-53]. Asteroids range in size from a few metres to many kilometres and their typical speed of travel around the Sun is 40,000 miles per hour. It is notable that many asteroids have rubble-strewn surfaces that look to be in a real mess. Contrast this with the much smoother surfaces we see on comets where it is obvious there is no great amount of debris lying around. This observation has a plausible electrical explanation if we again consider the 'scouring and cleaning effect' that comets undergo when they are subjected to plasma discharge action. Asteroids, not being significantly different in electrical charge from their environment, will not undergo this process akin to 'surface cleaning' so they tend to retain items of rubble of all sizes. Again, we have a logical, simple explanation.

**Meteoroids**: Think of these as being physically small and consisting of rock that often has a metallic content; they can even be as small as grains of sand. Meteoroids are precisely the same objects as Meteors and Meteorites, the only difference being 'where' they physically happen to be at any point in time. If 'the object' is out in space it is called a meteoroid; if it is passing through an atmosphere such as here on Earth then it is called a meteor; if it survives the journey through the atmosphere and hits the planet's surface, then we call it a meteorite.

Meteoroids seem to manifest within the solar system in a variety of ways. In terms of those we are aware of, some appear to have haphazard orbital patterns around the Sun with no obvious origin, others seem to travel predictably in groups and yet others due to their regularity, appear to be debris machined from comet surfaces when those bodies have previously entered the inner solar system on their predictable orbits.

A meteor fireball - Credit: Robert Mikaelyan

It is also possible for us to see that some meteoroids will have originated from planets and moons as their surfaces underwent catastrophic electric machining events in the past - here I think of the meteorite from Mars discovered in 1984 and labelled 'ALH84001' that led scientists to believe they had come across fossilised organic material that indicated life existed on Mars in times past [6-54]. The speed of these often tiny objects can vary from around 30,000mph to 200,000mph, and their collision speed with the Earth will vary greatly when we take into account the direction of travel of the Earth. This is where we have the Earth either moving into on-coming or relatively stationary meteor showers or moving away from showers that close in on us from behind our direction of travel. In both cases the speed of the Earth's orbit around the Sun, which is approximately 67,000mph, would need to be taken into consideration. What then about a typical meteoroid's electric charge?

Being much smaller than comets but fundamentally behaving in the same way electrically, we can think of them as tiny rocks that are only able to hold a limited amount of electric charge, which they can take on and release very rapidly. Because of this, meteoroids will always have relatively insignificant levels of stored energy due to their small mass. Being at the low end on the electric energy storage scale, therefore, they will be subject to levels of internal and surface electric stress that will not be significantly powerful. Nevertheless, being charged bodies that travel through Earth's ionosphere as they enter our atmosphere, their disintegration is often accompanied by bright electrical flashes.

When meteoroids become meteors as they enter Earth's atmosphere, they encounter ionised molecules of gas in the upper atmosphere that interact electrically and mechanically (kinetically) through friction to slow their small mass down and produce the spectacular displays we often see. Meteors are also what we call shooting or falling stars because we actually see their burn-up process taking place. If a particular meteor is large enough, part of it may survive this tortuous journey to strike the ground as a meteorite; these often having a black, electrically burnt appearance [6-55].

**Craters and Surface Scarring**: In terms of the craters we see on the surfaces of almost all solid bodies in the solar system, modern man has never observed any of these actually being produced. As was mentioned earlier in chapter three, the presumption that has evolved into accepted fact is that nearly all craters are formed by kinetic energy released when high-speed solid projectiles strike the surfaces of larger solid objects. Other theories too are accepted that explain the phenomenon of cratering, a couple of these being volcanic type eruptions and surface crust collapses. If any of these theories are valid, especially the impact theory, then is it not strange that nearly all the craters we see are round in shape? Should we not expect them to have other shapes as well, both regular and irregular, especially an elongated elliptical form that would surely result from projectiles striking surfaces at shallow angles? It seems no matter how hard one looks, irregular craters are very rare indeed and the circular variety seem to abound. In my view, it is wrong for astro-scientists to claim that nearly all craters have been formed through the impact theory. However, it is not wrong of them to claim that this process could indeed be one of the causes, albeit a minor one. The truth as I see it is that electric discharge between bodies is the far more likely cause of cratering. We will return shortly to this and other reasons to question impact theory.

The power delivered through plasma discharges that produce blemishes and features on the surfaces of bodies in space is a really scary level of power and a phenomenon not obvious to us today. We know that individual bodies in space can have a significant level of electric charge associated with them, so think again of when two bodies with vastly different charge levels come relatively close to one another within the conducting medium of space. Here the same situation exists as when a comet approaches the Sun and its surface erupts electrically to cast off material into space. Consider this process on the much larger scale where Jupiter and its moon Io are very close to one another. This is where we have observed what looks like erupting volcanoes on Io's surface. The similarity with comets is obvious [6-56].

Jupiter's moon Io and the 'erupting' volcano Loki
Credit: NASA JPL USGS

If one looks at what NASA astro-scientists say about this, on one hand they claim the ejected material is frozen sulphur dioxide that rains down as snow, but in the same article [6-57] they say that the temperature of these volcanoes is around 1800K - are they allowed to have it both ways? Also, rather mysteriously, we now know that some of those features on Io that have been interpreted as volcanoes have actually been observed to move around; what do you think might cause that?

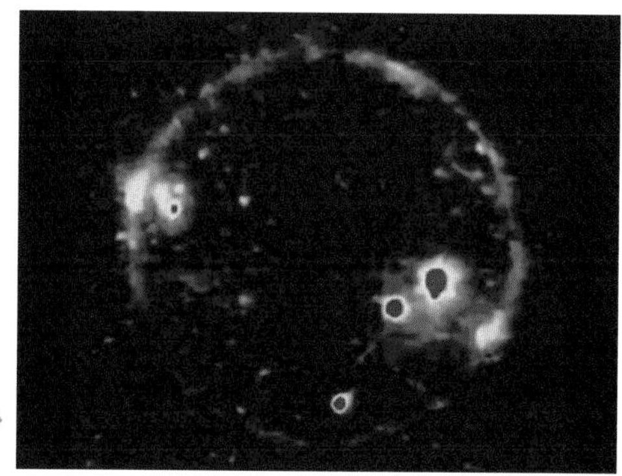

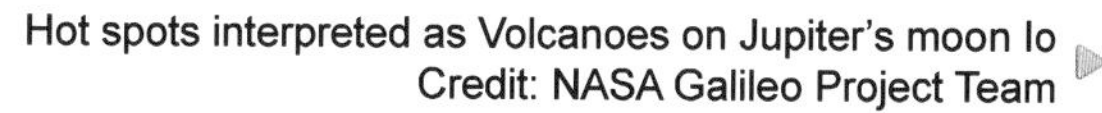

Hot spots interpreted as Volcanoes on Jupiter's moon Io
Credit: NASA Galileo Project Team

On what factual basis are we being asked to believe that Io's volcanoes are real and that they spew out very hot material close to 2,000K, onto the surface of that predominantly icy moon? Well, none, but the short answer we are being given is 'sub-surface friction' due to Jupiter's gravitational influence on its little companion. Because Io orbits close to Jupiter, it is assumed that despite it being a body with a surface temperature estimated to be around minus 150˚C, the tidal action of Jupiter's gravitational pull on Io's surface crust as it orbits the gas giant is a violent enough action to cause sub-surface friction to the extent that the material there is rapidly melted and ejected in massive plumes to a height of 300 kilometres.

Earthbound astro-geologists together with their terrestrial colleagues who are funded to come up with these ideas are known to seriously associate some of Io's features and observed activities to things they feel comfortable with from our environment here on Earth; things such as the geysers of Yellowstone National Park and their behaviour. I do not suggest this is totally irrelevant but it does highlight an interesting human tendency to first relate what is found to be puzzling to that which is familiar and comfortable. I leave this for you to ponder as to its relevance as an approach to forming scientific opinion that is typically closed to alternative ideas.

Similar in size to the Moon, here is little Io casting its shadow on giant Jupiter
Credit: NASA Cassini Project Team

EU theory gives a better explanation for Io's hot spots. Dark mode Birkeland currents exist between Jupiter and Io, where the surface touch-down points of these currents have mobility so traverse across the face of Io. The electrical machining on Io is therefore precisely the same as that on comet nuclei where the surface is etched by an EDM process on a cathode surface. The "volcanoes" on Io are merely scaled up cometary cathode jets.

**Geology thoughts**: I have looked closely at images and other data on craters and surface scarring on bodies in our solar system and am now well aware of the typical explanations put forward for their formation. In doing this I have also become aware of the distinctly defensive and occasionally arrogant posture adopted not just by the astro-science community in general but by many geologists and astro-geologists as well. I see this as an unhelpful and arcane attitude that has persisted too long but which unfortunately still continues. It is my opinion that the environment this has encouraged will be seen in future for what it has done to retard the healthy progress of terrestrial and solar system geology over many decades. Reasons behind this appear to include the well known fact that change is normally resisted, where this seems especially true with old scientists who remain with influence and who continue to work in the field of geology and astro-geology. I think that due to their now understandably limited earth-bound data sets and experiences, a great many of those fossils have convinced themselves, and unfortunately their students who believe every word they utter, that the central theories to which they hold fast are unquestionably correct and that any alternatives are a waste of time. Their steadfast positions seem further justified because they and their students can physically get their hands on many of Earth's geological features - features the origins of which they have convinced themselves they fully understand. It is however now the view that in cases involving such things as mountain chains, island formations, rock-strewn plains, mesas, valleys and canyons, that many of these display physical features that have much better alternative explanations that contradict standard geological theory, as it relates to Earth's accepted history. Despite this, arcane narrow views continue to be confidently delivered to all and sundry by the old traditional 'experts' in the fields of geology and astro-geology.

It seems to me that these people are at present and may forever be doomed to remain, unenlightened in terms of the fundamental electrical force and its structure forming and destroying potential. Although aware of the existence of electricity, as most people are, their thoughts never seem to include this important aspect. Instead, they wander among and within features of Earth's landscape and feel a closeness with those natural wonders, whilst also believing themselves to be experientially and theoretically well enough equipped to understand precisely how they and everything else came about. It appears plain that many of the assumptions and theories to which geologists hold have been arrived at to fit with old, general and simplistic geology models that are now shown to have highly questionable aspects to them. So what should those of us who feel it is right to demand more openness, cooperation and accurate information from Geology and all the other relevant '... ologies', do?

Credit: Author's Darling Wife, Nora

I suggest we should care more about what these and other disciplines produce for public consumption. At the same time, while recognising that not everyone wants to be or can become interested in these things, those of us who are motivated should continue on our journeys towards greater self-education on these matters. If we do, the path ahead will likely become clearer through involvement with modern theories and constructive debate.

As a more moderate comment, the most unfortunate aspect of geologists, astro-geologists and astro-scientists having their boots stuck in the molasses of arcane questionable theories while clutching tightly their half-empty tool boxes, is that established theories regarding Earth's geology have been and still are being transferred as also applicable to what we are discovering now on other bodies in our solar system. The collective report card of these undoubtedly well-intentioned people should therefore say, 'room for much improvement!'

Getting back to craters again, I previously mentioned that within our solar system the planets and their moons were previously not organised the way we see them today. There is substantial evidence that this is true and it will help to go with that idea for now so that an explanation of cratering and other surface features on planets and moons can be presented in context. To start, I will take you back to the previous reference to comet Tempel 1 and Wal Thornhill's predictions of electrical activity on and around it that thoroughly demolished the snowball model predictions of NASA scientists, which if acknowledged by them would have been highly embarrassing.

Comet Tempel 1 came towards the Sun from the outer reaches of the solar system endowed with its own significant negative charge. The copper projectile fired at its surface as part of the Deep Impact mission was positively charged due to its location at the time within the heliosphere close to the Sun. The voltage difference between the two bodies was therefore very significant and meant that before the projectile actually struck the comet's surface, an electrical discharge between the two bodies was certain to take place.

The 'strike' on Tempel 1 Credit: NASA JPL Caltech-UMD

This did happen and NASA scientists explained it away by claiming the comet had a thin, solid outer surface layer that the projectile must have come into contact with first. Furthermore, when the projectile hit the surface, the energy released threw up a tremendous amount of fine material in a spectacular show that was about ten times greater in quantity than NASA had predicted would come from an icy body. This more violent event was actually brought about by the significant potential difference between the material of the comet's nucleus and surrounding space, where just as fine dust is attracted to accumulate on positive surfaces in our homes, material from the comet's surface was ionised and attracted into positive space.

Just before the copper projectile struck the surface of Tempel 1 at more than 10 kilometres per second, some pictures were taken by its doomed on-board camera that were transmitted back to the Deep Impact probe. Those images revealed the surface of the comet to have two round craters.

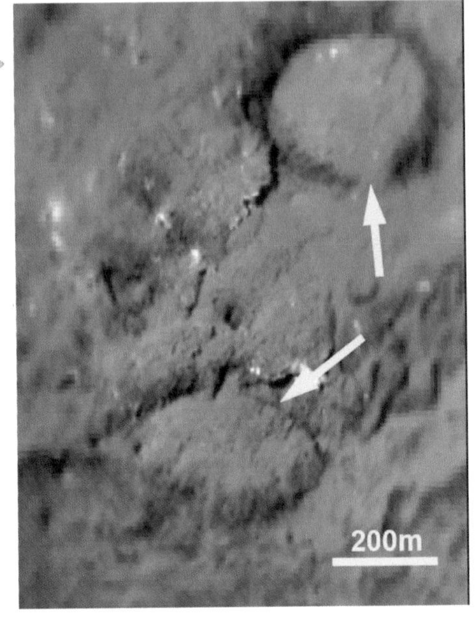


Tempel 1's 2 craters Credit: NASA/JPL-Caltech/University of Maryland/Cornell

This was the first time that astro-scientists had actually seen what they then swiftly assumed were 'impact craters' on the surface of a comet. Previous close-up images of both comet Halley and Borrelly had shown no craters, but interestingly, clear images of comet Wild 2 had revealed many 'circular depressions'. These were judged at the time as a puzzle hard to account for so a clear explanation was not forthcoming. Those depressions also had flat bottoms; that being a common feature of many craters found on the Moon, on Mars and on Mercury. The typical explanation for flat-bottomed craters is that after an impact event, molten lava from beneath the surface crust of the impacted body has oozed out, levelled off and cooled to form a smooth base. Consider this idea in terms of relatively tiny comets in cold space. It is close to impossible that they would ever have had lava under their surfaces, so what are we to make of this lava idea?

As is the case with the two craters found on Tempel 1, Wild 2's depressions are actually the result of electrical surface machining by the powerful material removal action of EDM. This happens in two ways: by close contact and subsequent catastrophic electrical discharge between bodies or slowly as less powerful discharge currents flow between the comet's nucleus and its oppositely charged space environment to erode material and leave sharply defined features. These are the fundamental processes at the root of most craters and surface scarring, no matter on what size of body they are found. Note in particular that in the images now available and because material is being removed directly to space, there is a lack of rocky debris lying around the locations of these electrical events.

We spoke earlier about planet and moon capture and the chaotic, close interplay that would have gone on between those bodies. Many of the planets, moons, comets and asteroids as individual charged bodies would have come close to one another. Discharge events between them would therefore have taken place that likely produced most of the cratering and surface scarring we see on these bodies today. Since then, similar electrical actions will also have taken place, they no doubt still continue and always will as comets pass through differentially charged regions of plasma and come close to oppositely charged bodies.

Impact theory does seem to be the obvious 'no-brainer' answer to cratering. However, with a logical approach and an awareness of the powerful electric force that exists in space, I suggest that among other good reasons, it is mechanically improbable and conveniently too simplistic! Looking further, there are terrestrial clues considered as acceptably analysed enough by astro-scientists and geologists for them to continue offering their analyses to the public as settled fact. An example of this is the 1 kilometre in diameter Barringer crater in Arizona, sometimes called 'Meteor Crater'.

Meteor Crater - Credit: Lunar and Planetary Institute

Despite the ludicrous sum (at the time) of 10 million dollars in the early 1900s being spent over 27 years in searching for the large and commercially attractive iron meteorite that was assumed had been the cause of it, no significant piece of that meteorite was ever found. Since then, and because of this failure, it has been assumed that the meteorite struck with such force it had been vaporised together with any significant pieces of rubble from the Earth's surface that might reasonably have been expected to be found still lying around the area. Yes, there are absolutely no large telltale pieces of rubble around the site of this crater, just a selection of fine debris. However, within this pebble and dust-sized material, significant amounts of a type of quartz have been found, and it just so happens that this quartz is the same as what we know is produced by lightning strikes on the ground. Moreover, the structure of the ground beneath the crater shows no sign of being greatly disturbed as one might reasonably expect it would be had it been the impact site of a large meteorite. In commercial and scientific terms, Barringer crater is an example of substantial effort being put in based on limited information and little relevant scientific guidance, the sum total of which was supported by assumption and wishful thinking that at the end of the day led nowhere useful for anyone. So much for how we can dedicate ourselves to what we wish to be true [6-58].

Around the time Barringer crater was being investigated, debate was going on in terms of the origin of craters on the Moon; this included both the 'scooped out' and 'flat-bottomed' types. The favoured notion was that they were the result of volcanic action, where in terms of the flat-bottomed types, an outflow of lava had settled and cooled to form a smooth surface. Even the great flat expanses of the 'Lunar Maria' (the Moon's so-called seas) were awarded this explanation. However, when astronauts eventually went there it was found very doubtful that the Moon ever had any kind of volcanic activity at all, so the 'volcanism idea' was changed to one where the impact of rocks had broken through the surface crust to release molten material that had been lying underneath.

It seems once again that what was known about the behaviour of volcanoes and lava here on Earth was transferred to what they thought they were seeing on the Moon. Unfortunately, this explanation has been stuck to ever since with no room for broader discussion on the matter. It is the line of thought automatically transferred to apply to similar features on other bodies, but it is especially the case with the subject of flat-bottomed craters [6-59]. Astro-scientists continue to cling to mechanical impact and/or geological disturbances as the only possible lines of explanation for craters and other surface scars. To me this is an example of self-defeating tunnel vision.

I also must address the claim from astro-science that round craters and their internal features can be accurately reproduced in the laboratory, no matter the angle at which a projectile is fired at a surface. Keep in mind that the obvious idea of craters being formed through impact events and the understandable human desire to be correct, would together likely form the backdrop to people's thinking as they considered why they should support that simplistic idea. I liken this to where a person with a narrowly developed belief will tend to ignore anything that contradicts it while recognising only that which is supportive. People just like to be right!

It is apparently the case that experiments have been carried out using miniature high velocity projectiles to show that round crater-like features can be produced no matter at what oblique angle the projectile is fired into sand or some other analogue of compacted surface-type materials. The results show simple crater-like shapes but none with the distinguishing external or internal features of real life craters. Arguments put forward based on this inadequate evidence should therefore be rejected, and in my opinion, those who conduct these expensive experiments should be put on the spot with questions about the accuracy of their results when compared against actual observations.

As Mr Spock of Star Trek fame might say to explain how the vast majority of craters are formed… 'Let logic dictate that due to evidence from direct observation, surface cratering is undoubtedly produced by a force acting vertically and in a rotating fashion to remove material quickly and cleanly from the body.' This really is what the evidence indicates, so to explain cratering and surface scarring in terms of EU theory we must concentrate completely on electrical discharge events [6-60] [6-61].

Experiments in electrical and plasma laboratories can now be done to reproduce major features of the cratering and scarring we find on planets, moons, comets and asteroids. Further to this, by-products of these experiments are frequently produced that are useful electrical clues, such as here where we find small round globes or 'spherules' known as Martian Blueberries.

Hematite Spherules (Martian Blueberries)
Left on Mars & Right from Laboratory
Credit: NASA (Left) and C.J. Ransom (Right)

Various fine sizes and quantities of these blue-grey 'Hematite balls' have been found in surface material lying around the perimeters of Martian craters. Geologists try to explain these through various mineral deposition, weathering and chemical process that are all tied to earth-bound reasoning, but it just so happens that these blueberries can be re-produced in the laboratory by exposing Hematite to powerful electric discharges [6-62]. The main experiments in this area have been conducted by the American Physicist Dr. C J Ransom, one example of which produced the result shown in the previous image.

The fact is that the vast majority of craters are circular due to being formed by a plasma discharge acting vertically on the surface of large solid bodies. Where significant and sustained current density is involved, the natural rotating action of the Birkeland current is what breaks up and scours out a depression that displays telltale indications of a powerful rotating action being involved.

Euler Crater on the Moon - Credit: NASA ▷

Take for instance Euler crater on the Moon. Here we see indications of a rotating action that has left a distinctive stepped spiral pattern around the crater's inside wall together with an isthmus connecting the inner bottom edge of the crater to its central peak which itself is a feature that also suggests a bifilar rotating force has been at work. Note the lack of significant debris around the crater and the light coloured powdered appearance of the immediate surrounding area. This does not resemble the aftermath of a powerful physical impact at all, and a vast number of other craters and the areas around them show similar evidence. I suggest that overall, we have tended unthinkingly to accept the standard explanations for cratering without feeling a need to look closely at the detail in front of our eyes.

Craters on Mercury, the Moon and Mars are said to have been formed mainly during a supposed event in ancient times that has come to be known as 'The Late Heavy Bombardment' [6-63A]. This is actually another fiction; it is an event dreamt up as an explanation for the strange pattern of cratering found on those bodies, especially on the Moon and Mars where one side of both these bodies is more heavily cratered than the other. Interestingly, the story around this fictitious event supplies no explanation as to why cratering on the Earth is found to be so different from cratering on the Moon, which of course is presumed to be its offspring. One would think that where two bodies have always been physically close to one another and one of these is bombarded by a shower of flying rocks that is of great enough extent to also affect Mercury and Mars at the same time, then the other body, in this case the Earth, would similarly have been affected, but this is not the case. Cratering on the Moon is definitely not like here on Earth so I suggest the 'Brothers Grimm' would have been impressed by this tale.

We can closely inspect many craters and have even had our hands on some of them. Should it not then be reasonable to expect that our ability to deduce their cause through common sense and reference to laboratory evidence might triumph? Sadly no, this does not seem to be the case, for the existence of electricity in space is denied and so the blinkers of gravitational and mechanical forces stubbornly still hold sway. Astro-scientists just refuse to consider the electrical theories that go well beyond those of gravity to provide many of the answers they seek. Their collective lack of education in electrical and plasma science is holding them and science back while also doing nothing to enhance the knowledge of the interested public. I expand further here on common features of craters and cratering that generally go unnoticed or are taken for granted without thinking.

*More on the question of, why do craters always appear round?*

Timocharis crater on the Moon - Credit NASA LRO

Experiments claimed to reproduce convincing circular craters no matter the angle of trajectory of the projectile go against the apparent logic of the situation where one would think that a shallow angle strike would result in an elongated surface scar. This idea seems to me to be reasonable and would appear at the very least to be a possibility, yes? No, astro-scientists do not place much weight on this obvious aspect and tend only to offer their standard impact, volcanism and surface collapse models to explain crater creation. One of their lines of explanation is that small, fast moving bodies on course to strike large bodies with atmospheres will heat up and explode at the surface with enough force to produce a circular crater directly beneath the blast. An assumption of this model is that none of the significant fragments of the incoming projectile would actually remain – i.e. the power of the explosion would be significant enough to ensure that the whole projectile is completely disintegrated.

It is also interesting to note that with impact being claimed as the cause, it could never be the case that the various materials of which rocky projectiles will consist would always react to impact and heating in the way typically described. Every object would have its own speed, angle of entry, material construction and chemical make-up that would ensure at least some of them were not completely destroyed with zero debris and no tell-tale signs left lying around on the surface. So what we are being asked to believe is that all projectiles no matter their size, composition, speed and angle of trajectory will disintegrate totally to leave round craters and no debris or other visual clues in the surrounding area. I leave you to ponder this!

In my opinion, supporters of the standard theories have boxed themselves into a corner and are essentially blind to the possibility that their ideas are wrong. They seem to know or care little about electrical effects because their professors are likewise ignorant on the subject. Yes, I understand that for some, aspects of what I say in this book will come across as arrogant and even unfounded. This is something I regret, but as an interested member of the public who has put the time and effort into understanding these things, it seems plain that impact theory can explain only a tiny fraction of craters, especially the circular ones that make up the vast majority!

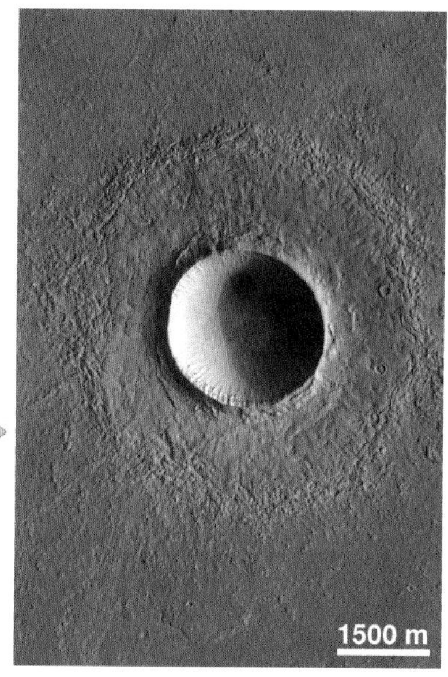


Martian crater on northern Elysium Planitia - NASA/courtesy of nasaimages.org

I see no option than to face up to the fact that if we accept that electrical currents flow within the conductive medium of space as described and that charged bodies exist there that have moved around and interacted electrically, then Birkeland current discharges will surely be the prime suspect in the process of machining out craters. This is also a process that people who have an appreciation of the basic behaviour of plasma discharges agree would take place at right angles to the surface at the receiving end of the discharge. Therefore, the conditions for producing a circular feature would appear to be in place. For visual confirmation of this just look at a decorative plasma ball. There you see every one of the dancing filaments of plasma in contact with the central electrode at right angles (90 degrees). So, to take everything so far covered into account, is it not plausible that a vertically-acting tornado of electric current with its associated EM forces could scour and lift surface material away from a planet or moon to leave behind circular craters?

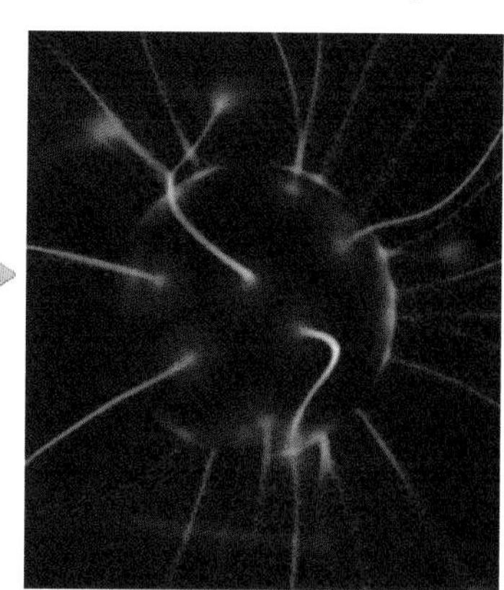

A Plasma Ball's 90° filaments © author

If you allow yourself to hold on to this thought then you may realise also that this electrical action would likely be completed in just a few seconds or minutes and that it would appear to an observer like the strike of a cosmic thunderbolt. Here again there is a temptation to link things with aspects of myth and legend found in ancient accounts of great battles that took place in the skies between gods and warriors. I find this profoundly interesting, especially because the science itself makes great sense and the peoples of those times would have had no reason whatsoever to mislead anyone with what certainly would be their honest attempts to record their impression of what was to them very real.

Consider the 'doctored' image on the right. It is intended to help with creating a mental image of the EDM process in action. Here the intention has been to show a twisting Birkeland current discharge to the surface of a solid body that is forming a crater by electrically pulverising and lifting material away into space as fine rubble that will likely end up as meteors and space dust.

Birkeland Current scouring out a Crater on the Moon
Original Image Credit: NASA

*Craters seem to often come in twos, where one is usually smaller than the other. This is a configuration found more commonly than chance can account for, especially as the smaller crater is often seen centred on or overlapping the raised rim of the larger crater, but never the other way around.*

Are we to believe that a large flying rock is often accompanied closely behind by a smaller rock that always strikes on or around the raised rim of the larger rock's crater? Of course we cannot believe this, but it is pretty much the stretch of imagination being asked of us when the detail here is considered! This configuration of craters can however be easily explained by electrical discharge, for lightning produces the exact same thing!

A plasma lightning bolt is made up of two or more consecutive strokes, both of which normally reach the ground. The fact that we typically see only one stroke is due to the speed and brightness of the plasma arc flash. The initial strike is the more powerful and subsequent strikes are less so. The detail of this can be seen in images of lightning strikes when captured on high speed film then played back in slow motion. Similar to this, plasma discharges to large solid bodies in space will first form a main crater then rapidly another smaller one as the follow-up discharge attaches itself to the highest physical area available; this typically being the raised edge of the first crater. This is what produces small craters on and around the edges of larger ones, a configuration that is common and one that in practice is never observed the other way around. Now that you know about this, I would lay money you will look for this crater arrangement when you inspect other images of planets and moons with dense cratering.

Moon craters - some in twos and threes - Credit NASA

*Internal crater walls are often seen to have a spiral-terraced appearance where this would not be an expected feature of an impact event.*

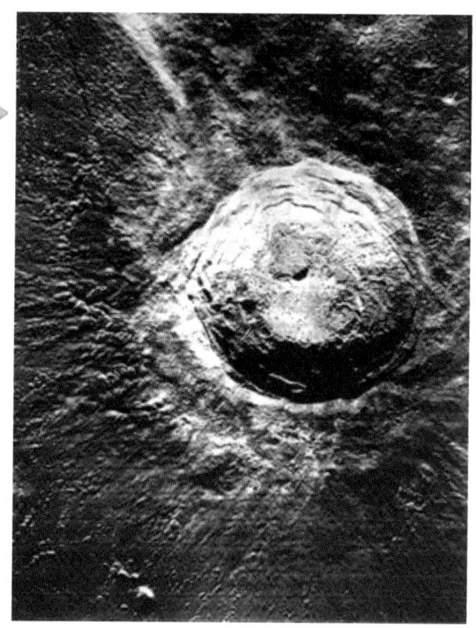

The crater Aristarchus on the Moon - Credit NASA - LRC

As we know Birkeland currents have a helical structure and a natural rotation about them. As a crater forming discharge event takes place the intensity (density) of the current will vary to some extent. The rotating scouring action that lifts material away will therefore be directly affected by those same variations in the power source. When this is taken together with what happens during the process of ionisation of the different surface materials and their various densities as they are being removed, it is possible to imagine a rotating plasma beam varying in intensity as its focus point darts about inside a crater to produce the semi-irregular but clearly recognisable internal spiral terracing that we actually find. Looking closely at detailed images of craters often reveals clues which indicate a rotating force that alters its focus point has indeed been involved.

*Why do we see small to tiny craters around the outside of some larger craters but not inside them?*

Where small craters are found outside and not inside larger craters that have the appearance of being freshly made it is an indication they were all formed at the same time. If the small craters had been formed before then they would appear covered by fine debris to some extent; they are not. And if they had been formed afterwards they would also appear inside the large crater; they do not. What therefore could account for this clear separation of crater sizes? A discharge event scouring out a large crater has subsidiary discharges of lower power taking place all around it that will also strike the surface. It is these lesser discharges that are the cause of the minor areas of cratering. See here an example of this with Lambert Crater on the Moon.

Lambert crater on the Moon. Tiny craters outside and none inside
Credit: NASA LRO

*The central peaks often seen inside craters are said to be produced by rebounding molten rock that has instantaneously 'frozen into place' after an impact event.*

Alignment of the strata of central peaks with those in crater walls © author

Standard theory claims that the central peaks of craters are formed by the instant solidification of rock that has become molten after a projectile strikes and heats an area on the solid surface of a planet or moon. The event being suggested here can be imagined as something akin to when you drop a stone into water and watch a central column shoot up. Again I share the view of others and cannot believe this is how these central peaks are formed. Instead, it seems logical to believe that a bifilar Birkeland current with its rotating machining action around a crater's central point could leave behind a single or tight group of protruding cone-shaped features. In support of this it is claimed by some that these peaks have the same vertical sequence of stratification (distribution of layers of material) that the inner wall of its crater has. In any case, a process for central peak formation supports the idea of a rotating scouring action that removes material and does not support the idea of a catastrophic impact event.

*We never seem to see any great amounts of rubble lying around the site of craters, even though this would be a logical expectation resulting from an explosive impact.*

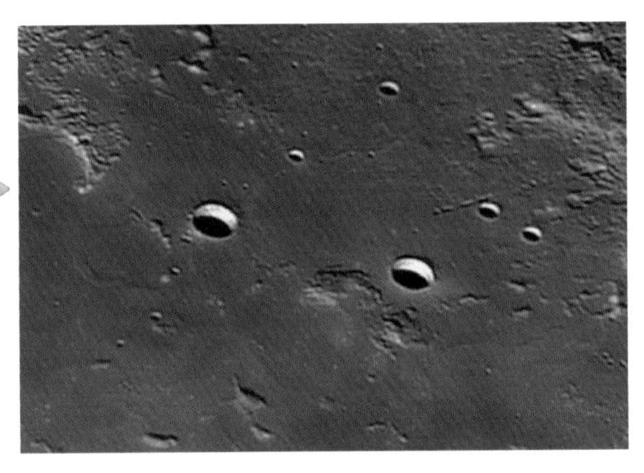

Why do we not see debris around these craters on the Moon?
Credit NASA LRO

It is practically impossible to find extensive rubble-strewn areas in the immediate vicinity of craters. Even where we see craters adjacent to one another, no material from one (referred to as ejecta) can be found lying inside another, so where has that excavated material gone? Due once more to the scouring action of Birkeland currents two main things have happened: most of the pulverised debris has been lifted upward into space, and the vast fields of sharp rock fragments, for example, those now found extensively across the surface of Mars, must be a result of much of that debris having fallen back to the surface after being lifted skyward. This is also the process that supplies us with an explanation as to why we still find small pieces of Mars as meteorites lying around here on Earth.

*Crater chains are common, where if these were produced by impacts then it is also to be expected that tight formations of flying rocks of similar size are at this moment travelling at speed through space in formations similar to that of the carriages of a train.*

Crater Chain on the Moon - Credit NASA ▷

A plasma discharge moving across a planet or moon's surface will tend on occasion to hesitate for short periods as it progresses. Again, for a measure of visual confirmation you can look at the often-hesitant performance of plasma filaments inside a decorative plasma ball. One result of this discharge behaviour on a body's surface will be to produce a line of adjacent craters, the rims of which might touch or overlap. Formations such as these are just impossible to reproduce through impact theory. However, if impact was to be considered, then we should expect to see one crater being physically affected by the creation of the next but evidence for this is not being found. On some occasions, we even see crater chains that change direction to leave geometric patterns that the theory of flying rocks could never reproduce.

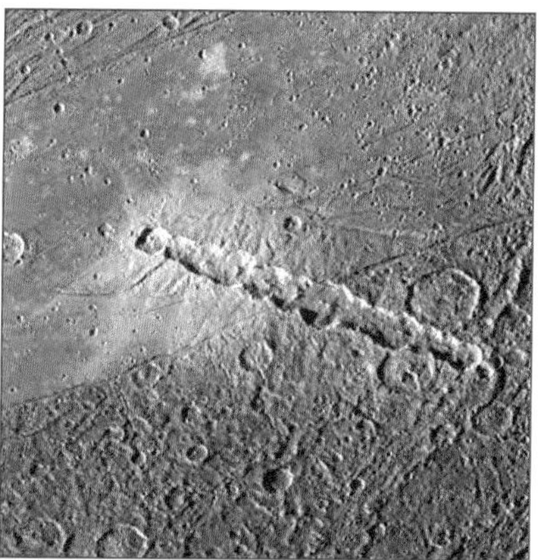

Crater Chain on Jupiter's Moon Ganymede - Credit NASA ▷

We should accept that a large body can indeed break up due to the action of electrical forces inside it being influenced by the environment of charged space to fracture that body and that a formation of large pieces might be the result. This is what happened with comet Shoemaker-Levy 9 as it began to pass Jupiter. That comet was however particularly large and its remnant fragments rapidly drifted apart as they neared Jupiter's atmosphere and were drawn in by its gravity. So basically, the idea of rock fragments remaining together in tight formation to produce the results we see in crater chains is not supported by the basic mechanics involved or by observation.

Crater Chain on Mercury - Credit NASA - Messenger Mission ▷

Any theory to explain crater chains must involve a relatively constant force that moves across a body's surface. In the case of crater chains the energy will likely be discharged in rapid bursts due to surface material densities and fluctuations in the discharge energy available from the source.

*Vast areas of the hemispheres of planets and moons show stark differences in the amounts of cratering they have.*

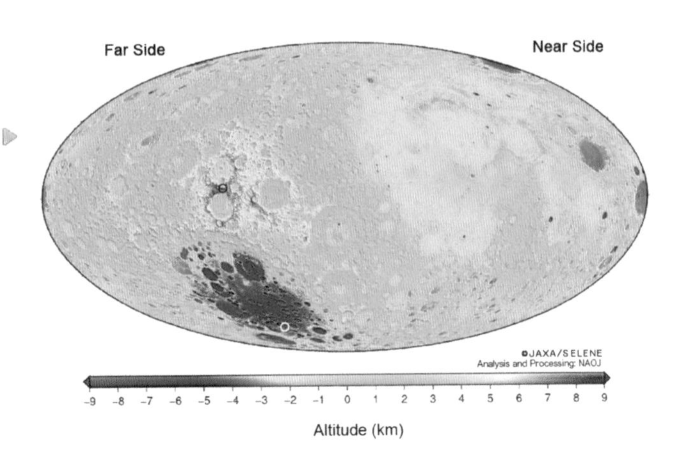


Both sides of the Moon showing cratering and altitude
Credit: JAXA - SELENE

This rather noticeable phenomenon is especially apparent on the Moon and on Mars. Since both these bodies rotate, any meteor shower to account for this strange distribution of cratering would need to have arrived rapidly all at once and from a single direction. I have already mentioned that relatively small rocks in space will not congregate in dense showers, so together with our earlier dismissal of a 'Late Bombardment' type event, plus the absence of any other serious explanation, this notion goes beyond highly unlikely into the fanciful. Instead, it is much more plausible that a 'large area plasma discharge' is responsible - this being a torrent of relatively small plasma discharges all acting on the surface of a body at the same time and probably emanating from another closely passing and differentially charged large body.

The idea of Birkeland currents extending through space between bodies as discharge currents is significant. However, it is supported by facts such as there always being available the medium of plasma in space for currents to flow through and that the power those currents are able to transport being considerable. These highly energetic events would likely be triggered between planetary and moon sized bodies when their respective magnetospheres/ plasmaspheres meet, whereupon an electrical imbalance would be established and forces to mitigate that state of imbalance would take over. In the image of Mars shown here we see a stark difference between the north and south hemispheres where the relatively smooth and uncratered north has had a great amount of material removed to leave an average elevation (in blue) that is significantly lower than that of the cratered and rubble-strewn south (in red). What could have removed such an amount of material from only one side of Mars?

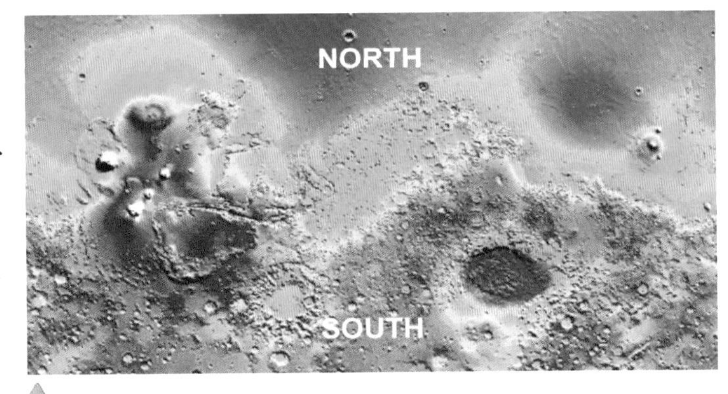


Mars' vastly different hemispheres - Credit: NASA/JPL MOLA

In considering further examples of surface discharges, there is the case of the famous crater on the Moon known as Tycho. The pattern of 'dust rays' around that crater and the rays themselves have been a source of long-standing puzzlement. Unlike what one might expect with large amounts of material being blasted out from a central point then falling back to the surface to leave evidence of heavy rocks having impacted the surrounding area, the rays instead seem to lie across the surfaces of craters and other features as light coverings of fine dust.

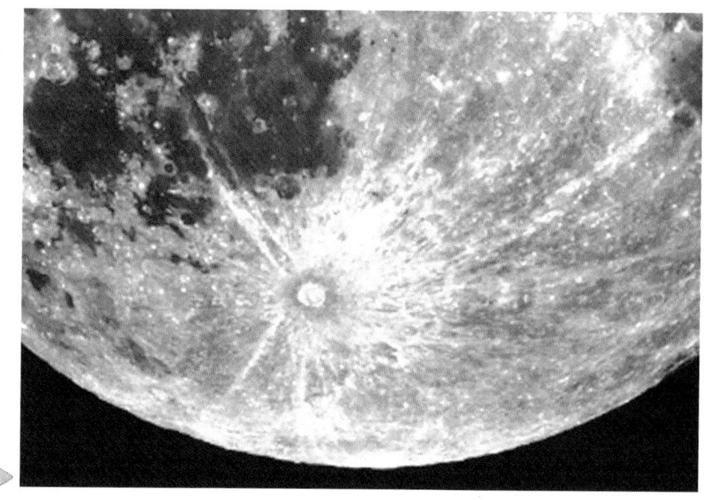

The Moon's crater Tycho and its 'Rays' - Credit NASA HST

Further to this, the rays extend far around the lunar surface while remaining very well defined, so much so that the Moon's own puny gravity cannot be used as a source of explanation for the extensive well-defined pattern they make. It seems that the reason behind these rays appearing as they do will remain a problem to solve if electrical discharge as the cause is not considered [6-63B]. (Ralph Juergens made a particular study of Tycho's rays that can be found here: http://saturniancosmology.org/juergensb.htm)

When everything is taken into account, the notion of magma oozing up and out over a surface after an impact event to form craters, many of which seem to have steep scalloped edges and flat floors, just does not cut it as an explanation these days. Once again plasma discharge can explain what we see when we consider that it has the ability to gradually erode away areas of material over periods of time, as was previously mentioned happens on the surfaces of comets to produce their sharply outlined features. This erosion involves relatively low but focussed current densities because it operates on small areas through plasma's dark and glow modes.

The image on the right of previously formed craters has been altered slightly by adding glowing edges to the craters in an attempt to illustrate the erosion effect which subsequently takes place. This 'electrical eating-away' of surface material is a slow activity that would naturally concentrate on raised areas and sharp edges. This brings to mind once more the mental picture where the ash of smouldering paper glows and falls away when one blows on it.

The action of electrical erosion on craters (Conceptual) - Original Image Credit NASA

How is it that the small bodies of comets and asteroids often sport a great number of craters and other 'impact credited' features on their surfaces? If their own gravity were to be offered in answer as the attracting force then how in terms of a comet's or asteroid's negligible level of gravity could it attract significant sized pieces of debris? No, this could never be. There is also the more popular idea of random collisions but even with this the probability of it happening with any great frequency and with the required force to form often outrageously large craters, is extremely unlikely, so this looks like it is not the answer either.

The cyclic passage of comets and asteroids through the solar system over great swathes of time will undoubtedly on occasion have brought them into relatively close contact with substantially larger and much more powerfully charged bodies, especially in the early days after comets and asteroids were produced as the debris of major catastrophic events. Think back again to the electrical forces in play that were described as responsible for breaking up comet Shoemaker-Levy 9 when its course took it a bit too close to Jupiter. The current density that comet experienced from its environment will have increased in intensity as it neared Jupiter. The result being a build-up of electrical stress between regions of isolated charge inside the comet and discharges taking place between those regions like internal lightning bolts that would fracture to cause the break-up of the comet. So with powerful electrical forces available to act on roaming bodies, we have an alternative to impact theory, especially since the size of some craters suggests that the struck body would have been smashed to pieces. Consider this image of the crater 'Stickney' on Mars' moon Phobos. One can imagine that the impact power required to produce this would never have allowed poor little Phobos to survive in one piece.

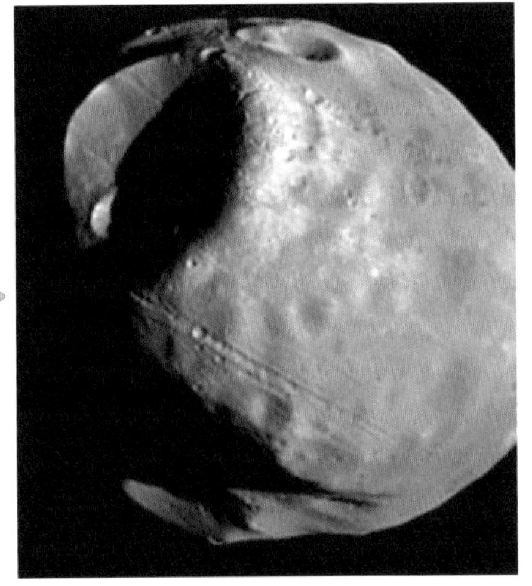

Mars' moon Phobos and its crater 'Stickney' - Credit NASA Viking 1 Orbiter

Here is a little more on the pointless attempt to use gravity to explain these craters. Due to its observed behaviour as it orbits Mars, Phobos is believed to be a very low density object, along the lines of imagining it is made of polystyrene. The real explanation for this apparent low mass was previously given in this chapter where a body's charge level and apparent mass were linked. For astro-scientists, however, this apparent extreme lightness of Phobos has been used as a basis from which to explain how that body would be able to absorb the force of an impact that would result in a crater the size of Stickney while the rest of Phobos stayed together. This is a blatant example of a sequence of assumptions relying on one another; not a good basis on which to pursue the truth of any matter! Phobos is instead a dense object held in place solely by Mars' own gravity. Stickney crater and the uncluttered area around it could only have resulted in the past from electrical interaction with a larger body and a subsequent EDM process taking place on its surface.

Other surface features that must be mentioned are the long and deep gouges (rilles) that appear cut into planet and moon surfaces. The EDM action that forms craters applies equally to the process of rille formation. The difference is that instead of a discharge event being focussed on a single area, lateral surface motion of the discharge event has to be considered as well. Mentioning rilles allows me to move on to the subject of 'surface scarring' and the creation of other features on planets and moons [6-64] [6-65] [6-66] [6-67] [6-68] [6-69].

**Geology, surface scarring and other features**: *Most of the surface scarring and other natural and unnatural geological features on Earth and other bodies are the result of plasma discharge events.* This statement is not one the public is likely to pay much attention to because we have traditionally taken the explanations of geologists and other science discipline experts as being sound because they do come from 'experts', so we just accept them as fact. Please think about this carefully for a minute. Earth scientists in general are not specifically educated in astronomy and are likely to have precious little knowledge of electrical or plasma science. Therefore as a basis on which to form their opinions, many of them rely unwaveringly on the traditional geological theories passed down to them in the academic setting by their teachers. This means that in the public arena as knowledgeable voices to be heeded, they feel within their professional and personal comfort zones and believe that the understanding they have of our Earth, which extends of course to include other bodies, is at best absolutely correct or at worst heading towards the truth. I intend no disrespect to the work and dedication of these people but when certain antiquated geology theories are passed down through a comfortable academic hierarchy to be used together with that discipline's lack of off-planet experience, then it seems clear that geologists and astro-geologists will prefer to explain things using a traditional toolbox that is actually not as well populated as it might be and which applies mainly within Earth's isolated biosphere.

To contrast this, when a broader modern view that includes the theories behind the EU model is adopted, then a very different vista opens up for consideration. Of course geologists have a stake in defending the fundamental theories to which their careers have been devoted; that is understandable. They also have every right to hold opinions on anything they choose, just like the rest of us. However, it does seem a pity that as theories directly applicable to their own discipline have made great strides, they have chosen to stand still and not consider an update to their own. My view of this and other related things is not an academic one. I see it rather as one of common sense based on a level of familiarity with the broad range of evidence available to all of us if we choose to open our eyes and learn more. With an open mind, feet on the floor and with a modicum of scepticism, one needs only to study some of the basics of geology and look at the compositions and formations of features around us to become suspicious of the traditional explanations that movement of the Earth's crust and water and wind erosion alone have been responsible for them all. This thought brings to mind physical features such as the Arizona Bluffs in the US where it is water and wind erosion that are said to have formed them. It is my now firm belief in the ubiquity of powerful electrical forces that leads me to say I cannot accept this idea.

Monument Valley Arizona - Source Wikitravel

In terms of the Earth in its dim and distant past it is suggested that because our environment in space was not then what it is today, things appeared very different in the skies and on the ground. This comes to us from the work of respected scholars of ancient history and comparative mythology who say that other bodies within our near space environment moved close enough to each other in those days for discharge events to occur between them that brought about surface scarring and general geological upheaval. This was the landscape sculpting that formed the valleys, mountains, volcanoes, bluffs, basins and other features on these bodies that today is attributed to antiquated theories. Once the 'Pandora's Box' of updated, inclusive geology is opened I think many will find reason to see the old theories under a new light and go on to form a sceptical view of currently accepted earth history. The insistence that large scale crustal movements, earthquakes, volcanism and wind and water erosion can explain just about everything goes too far with what is now understood by many to be limited information in the hands of the trusted but inadequately informed.

**The Grand Canyon** – We are told that the Colorado River once flowed with such force and volume over a period of time that it formed the fantastic Grand Canyon in Arizona, USA. This is quite strange because there happens to be no identifiable delta or other outlet to the sea or to anywhere else that one would expect to find if a cataclysmic flow of water was involved. Neither are there any great amounts of debris to be found anywhere that would have been washed out as the original material that filled that vast chasm. There is even an area of the canyon called the 'Kaibab Upwarp' where if indeed water had caused it then it could only have done so by running uphill. Furthermore, the question exists - where would the water have come from anyway?

The Grand Canyon from Space - Credit Rick Searfoss Retired Space Shuttle Commander

Debate still goes on around these and other things and for enlightened geologists the questions that arise are far from being settled. With the background of what we now know about plasma discharge ubiquity and power it seems worth considering that the Grand Canyon was machined out of the Earth's surface by an EDM type event in the distant past. After this event it is likely that its then freshly exposed surfaces were indeed eroded to some small degree by wind and rain. One wonders what today's geologists would say if evidence that electrical action had taken place was found.

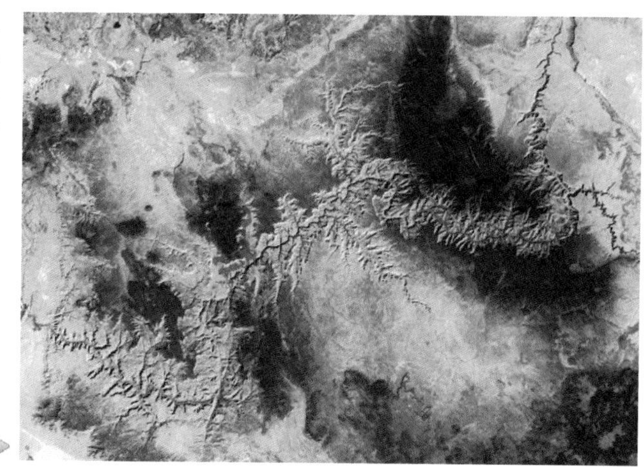

Grand Canyon from space - Credit: ESA European Space Agency

Take a really close look at the intricate pattern of the Grand Canyon in the pictures provided to see the network of main branches and the hierarchy of smaller branches running from them. Does this remind you of a pattern that you may have seen in other places? For the point I am attempting to make here, I can tell you that this is the pattern that an electrical discharge will produce both internally and on the surface of a body. It is called a 'Lichtenberg Figure' and an example produced in a laboratory is shown clearly in this image [6-70A].

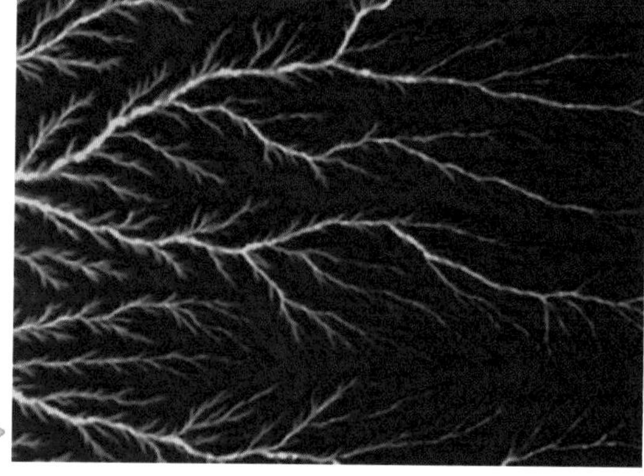

Lichtenberg Figure produced by Electric Discharge
Credit: Dr John W Gudenas

The association to be drawn here is highly significant, for we can see this same pattern being repeated in many geological formations on solid bodies in our solar system and also in many locations on Earth, such as the Grand Canyon. Lichtenberg type figures can also be found in a wide variety of locations across Earth from the deep ocean floor to vast open plains and to mountain tops. If you look at detailed satellite imagery you will see many examples that would traditionally be described as the results of water flowing from melting glaciers or widespread powerful flooding in ancient times. Some of this will of course be true but can we honestly put all the Lichtenberg figure patterns on Earth down to these causes? Please understand that this is not an attempt to dismiss good theory, it is rather a reminder to look closely and to make up one's own mind in the knowledge that the science behind these matters is not settled as a done deal. Now we will look at Valles Marineris on Mars.

The similarities between this chasm on Mars the 'Red Planet' and our image of the Grand Canyon are striking. All of the main features seem to be alike in so many respects; the steep sides and flat floor areas, the complicated major and minor gully systems and the meandering appearance of some branches that look suspiciously like the results of a lightning strike [6-70B].

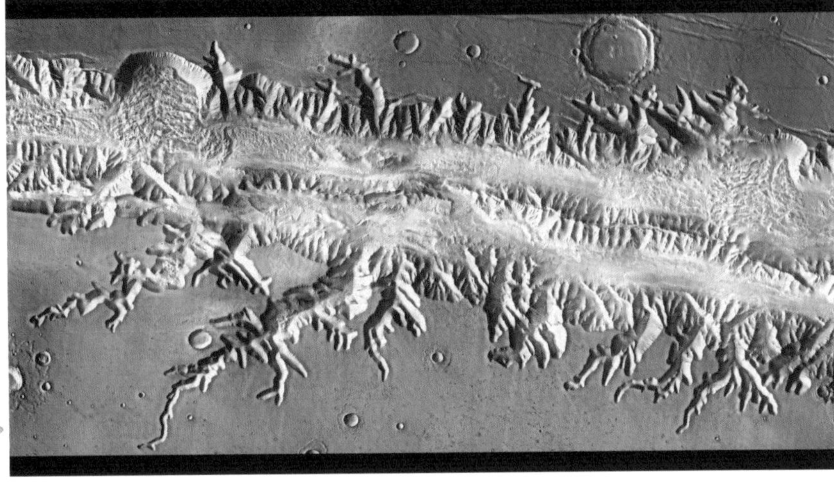

Valles Marineris on Mars
Credit NASA/JPL MRO

According to the standard view, Mars is a dry and cold planet that has abundant water trapped in a frozen state under its surface and especially at its poles. Further speculation is that any substantial atmosphere it may have had in the past was lost to space through forces yet to be explained satisfactorily by astro-science. Looking at the surface of Mars today you would be right to come up with questions about what we see there; crisp clean features that often look highly unusual and as if they were formed not too long ago. There are many deep channels in the surface of Mars of which Valles Marineris is the best known. Just like other much smaller, but just as cleanly formed, gouges that look as if they have been neatly machined out of Mars' surface, Valles Marineris abounds with clues that scream out to us that it was not formed by flowing liquid or by surface crust displacement or collapse. All these things are puzzling and many assumptions have been made in the attempts put forward to explain them. However, the tendency to guess at things has backfired and highlighted the fact that what works for Earth cannot be transferred to Mars and be expected to work similarly there.

Things become much clearer if we consider that most of Mars' features are the result of tremendous plasma discharge events originating from space, perhaps in the not so distant past. We have already mentioned the similarity of shapes produced, especially around the edges of Valles Marineris, with that of lightning's tell-tale Lichtenberg pattern. On close inspection these edges show neatly scooped-out features that have a distinctly repeated scalloped form to them. These clean scalloped edges and the various scales of Lichtenberg patterns on the inside walls of Valles Marineris provide significant indicative evidence that counters the claim that a great volume of liquid flow in the past or that some form of crustal splitting has been responsible for the formation of that gargantuan scar. There is no supporting evidence for seismic upheaval on Mars ever having taken place in the past so it seems that Valles Marineris cannot be explained through earth-bound geological theories and that EDM action is once more the likely cause.

Following is a selection of extracts from the TPOD article *'Message of Valles Marineris'* [6-71]

***'The greatest canyon in the solar system, Valles Marineris on Mars, underscores the contrast between two interpretations of the Planet's history. Now, high-resolution images of the chasm cast new doubts on old explanations.***

*In recent years, no Planet (apart from Earth) has received more scrutiny than our neighbor Mars. The 'Planet of a thousand mysteries' is more than an unusual member of the solar system. It has emerged as a laboratory in space for the exploration of solar system history. And the story it has to tell is so different from the things we learned in school that a retreat from all prior doctrines is now essential. Current geologic concepts, based on terrestrial observations of volcanism, erosion, and shifting surfaces, fail to account for the features of Mars, and the history and geology of Mars that have been built on those concepts is incomprehensible. But letting go of a cherished belief system often requires a shock.*

*Fittingly, it is the electrical viewpoint that provides the required 'shock to the system'. The contributors to this page believe that on the objective test of 'predictive ability' – the only legitimate test in the theoretical sciences – the electrical hypothesis will account for the dominant features on Mars, where popular theory fails.*

*Often the simplest test of a new approach is to consider its most extraordinary claim. Of all the enigmatic features on Mars, none is more striking than Valles Marineris, the great trench cutting across more than 3000 miles of the Martian surface. In our Picture of the Day for April 08, 2005 'The Thunderbolt that Changed the Face of Mars', we suggested that Valles Marineris was created within minutes or hours by a giant electric arc sweeping across the surface of Mars. Rock and soil were lifted into space and some fell back to create the great, strewn fields of boulders first seen by the Viking and Pathfinder landers'.*

*But what will it take for Planetary scientists to consider a new way of seeing Valles Marineris? It will require a willingness to reconsider all assumptions, without prejudice. A prejudice is an unfounded assumption that leaves one in a state of partial blindness. On the matter of Martian history in general, and Valles Marineris in particular, the most powerful prejudice is an untested supposition, the bane of space age science: the idea that Planets have moved on their present courses for billions of years. No one should have the intellectual privilege of asserting such an idea as dogma. The idea originated as a guess and then, in the absence of any definitive evidence, crystallized into a doctrine held in place only by the inertia of belief.*

*The second requirement is to allow for the possibility that the Sun and Planets are charged bodies so that, within an unstable solar system, electrical arcing between these bodies may have been the dominant force that carved surface features. Yes, this is an extraordinary possibility, but it is also supported by an immense library of evidence, as we intend to show in these Pictures of the Day.' 'For a time, the most plausible instance of surface spreading was Labyrinthus Noctis'...'Some scientists had compared this region to the cracked surface of a loaf of bread as the surface is raised and spread during baking.'* [6-72]

*'As seen in numerous counterparts on Mars, the depressions of Labyrinthus Noctis appear as complexes of crater chains and flat valleys, cut by the same force that created the overlapping craters elsewhere on Mars. The surface areas untouched by the arc thus remain as buttes and surrounding plains above scalloped cliffs. The smooth surfaces above the valleys show no evidence of rifting or of the supposed stresses that are claimed to have 'torn' the surface, just a complex of even more shallow, flat-bottomed and often parallel grooves, a recognized signature of electric arcing.'*

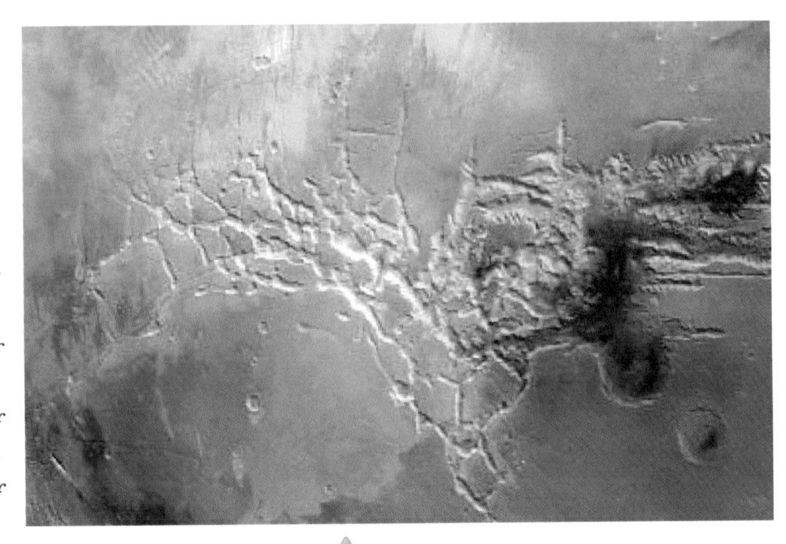

Labyrinthus Noctis on Mars - Credit NASA/JPL

**Rilles**: These surface gouges are to be seen all over Mars, Mercury and the Moon. They are also found on other moons around planets in the solar system including Ariel, Titania, Miranda, Triton, Enceladus and others, where in every case the surfaces of these bodies seem again to have been subjected to the effects of electric discharge. There is an abundance of TPOD articles available on the thunderbolts project website at www.thunderbolts.info that describe in very readable detail examples of the types of surface markings on solar system objects to which I am referring here. I leave these to your own further studies and will concentrate now on the rilles to be found on Mars.

If you dig the tip of a spoon into ice-cream and draw it along to scoop out a channel then you have formed the same simple shape that a rille has. However, the real thing would of course have its dimensions of length and depth in kilometres. You would also note that no debris is left lying around the edges of the gouge because you cleanly removed that tasty 'material'.

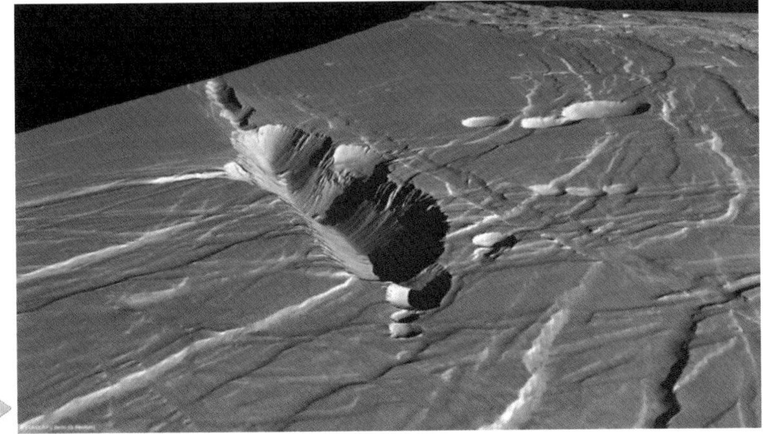

Phoenicis Laqcus Rille on Mars
Credit ESA/DLR/FU Berlin (G. Neukum)

In terms of the rilles on a large body like Mars, material would first be 'electrically pulverised' by a traversing discharge event then drawn upward from the surface as all grades of debris, some of which would fall back to the planet to form rocky debris fields and some of which would be carried on into space as wandering meteoroids.

Echus Chasma (1) Rille on Mars
Credit ESA/DLR/FU Berlin (G. Neukum)

One need only consider basic physics and geology concepts while studying the now widely available detailed images of Mars' surface to see that flowing liquid, surface crust movements and the assumed collapse of self-supporting kilometres wide lava tubes could never have produced the complex, smoothly formed and debris free rilles that often crisscross each other and indicate that if liquid were involved then it would have needed to flow in ways that defy basic physics. It seems clear instead that a traversing plasma discharge has been responsible for gouging rilles and for producing other significant surface features as well.

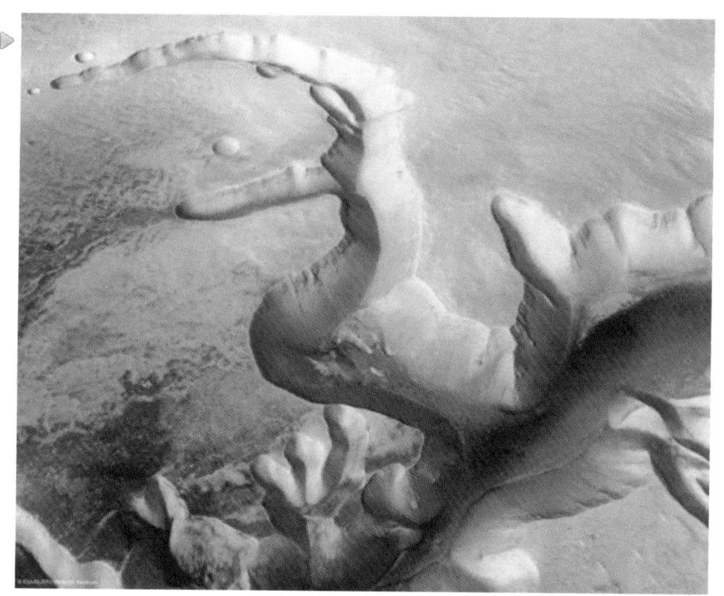

Echus Chasma (2) Rille on Mars
Credit ESA/DLR/FU Berlin (G. Neukum)

It is somewhat ironic that we have until now had NASA and ESA as the main originators and distributors of high-resolution imagery returned from their MRO (Mars Reconnaissance Orbiter) and Mars Express missions. These organisations are committed to releasing (usually slowly) imagery sent back from these missions into the public domain, but are at the same time the best known public adherents to standard theories. Detailed inspection of images by independent experts may then be the activity from which the abundant clues being revealed go on to prove beyond doubt how rilles and other Martian surface features have actually been formed.

It may therefore be NASA and ESA themselves that hold the key to the demolition of important aspects of long-held cherished theories [6-73] [6-74] [6-75] [6-76] [6-77] [6-78] [6-79] [6-80].

Much more could be said about craters, surface scarring, rilles, anomalous other features and the processes that form these things. This, together with the more scientifically elaborated detail of EU theory is better left to those plasma and electrical scientists, technical professionals, authors and otherwise qualified individuals who can do a much better job of describing them in scientifically sound detail. In addition to the links I have provided I would implore individuals reading this, those who are motivated to look further, to make a point of visiting the main web site relating to the Thunderbolts Project and the other main sites related to it - these are:

www.thunderbolts.info - www.holoscience.com - http://electric-cosmos.org/ and http://astrotes.info/

---

To wind this chapter up, here is a further selection of images of the Grand Canyon and areas around it, which in terms of the scepticism I now hold for tired old geological theories, give me cause to consider how many of these features were actually formed. With my now firm belief in the legitimacy of the theories that support the EU model, I cannot help but compare what I see in these pictures with what we are now able to inspect in great detail on the surface of Mars and in far clearer satellite imagery of Earth than was available before. I truly wonder if evidence of powerful electrical discharge events in the past is right in front of us here.

If Earth's geology is of particular interest, then you may want to look further into some of the alternative ideas that exist around it. There is one researcher that I would mention in particular, Michael Steinbacher. Michael devotes much of his life to exploring, analysing and explaining from a very interesting viewpoint, the multitude of geological formations we find here on Earth. His concept of how things have come to be as they are diverges substantially in a thought provoking way from the mainstream to include aspects of Electric Universe theory. When Michael's work is considered together with what I have tried to describe in this book it does make a good deal of sense to me. You can find a paper and presentation about Michael's work here [6-81] [6-82]

---

(The following pictures are provided by courtesy of James O'Sullivan - www.planetjim.com)

I find these and many other pictures of the Grand Canyon, remarkable and a great source of food for thought.

The questions become obvious when the Grand Canyon is considered in its complete form. Why are there so many sharp edges and turns in the gullies? Where are the inlets and outlets for the water flow that is supposed to have taken place? Where would the vast amount of sudden water flow have come from to start with? If water was indeed involved, then where has all the washed-out debris gone? Why do we see everywhere the Lichtenberg-type figures that are also associated with electrical discharge events?

A broad, open basin like this is more representative of a lake bed than the bed of a supposed fast flowing river. What would cause such a huge volume of material to be removed?

These gullies have an almost erratic branching form that does not seem reminiscent of what a powerful flow of water over a long period of time would tend to do as it cut through solid rock.

Would a large volume of water flowing over time leave a nicely defined V-shaped slope down to the river bed? I personally do not think so.

Navajo Canyon

What could cut this pattern out of solid rock? Could it possibly have been a traversing, twisting force?

Slot Canyon

Lichtenberg figures are once again obvious in the patterns that have been produced. It seems here that layers of strata have been removed vertically and not horizontally as one might expect to be the case with water flow.

## Flat Top Mesa

It may just be me, but this solitary rock outcrop seems to scream out that material has been removed up and away from around it and not as a result of anything like water flowing past it.

## Rockballs

These lie around in the vicinity of the Grand Canyon and they remind me of the so-called 'blueberries' found on Mars. Similar objects can be reproduced here on Earth by directing an electrical discharge into Hematite.

# 7 | A final word

Back in the good old days the whole of science was called natural philosophy. That was a time when educated people curious about nature and in what might exist outside our world, talked together with little restraint to exchange their ideas. Theirs was a world of investigation, collaboration and discovery that must have been a wonderful experience to be part of. I choose to believe those natural philosophers considered their contributions to be elements of what they were naturally meant to do in life. From this I further take the view that fair-minded folks today might think that professional involvement in any aspect of science should be seen by its participants as a vocation that has a moral code to be followed together with a duty of responsibility to unhindered human advancement. In a perfect world, science would be beyond reproach.

Because things are not like this today, how can we still make some difference for the better? I could not reasonably expect to hand this book to just anyone on the street and expect them to understand its content, its reason for being, or even to have a hint of interest in the subjects involved. Instead, it would be best received by those members of the public who are already actively or potentially curious about science and who feel that for self advancement they must add to their knowledge and connection with our natural world.

However, many important things are happening today that affect us as individuals and it takes a stretch of cold reality to consider that the story of our universe should be more important than finding or keeping a job, paying the mortgage or feeding the family. I think, therefore, that since the question of how our universe works remains an academic one, in that it does not affect our everyday lives in obvious ways, it will understandably remain secondary to the things that are currently more fundamental to our needs and responsibilities.

Nevertheless, we humans need to maintain some level of mental stimulation if we are to remain balanced and on-track to develop properly, so perhaps this aspect alone is a good enough reason to pay attention to this broad subject. If enough people in the public arena are seen to understand the nature of the questions that astro-science has for so long been able to fudge answers to and fob off because there has been no penalty for doing so, then perhaps mainstream scientists will begin to realise that public opinion is important and that in terms of where science is headed, keeping that audience properly updated with good information is the right thing to do.

This has been an introduction aimed at the interested non-professional who may or may not have already heard about the Electric Universe model and its theories. It has hopefully been a joined-up summary of the basic ideas behind those theories and an attempt at highlighting the inferior nature of the gravity-related explanations we have already been given for the formation and operation of everything in our universe. What is being suggested as the alternative science that applies throughout the universe is significant and important for us all, and my hope is that the reader has been able to see for themselves some justification for the case that has been made.

In terms of the job I originally set myself, I saw it from the start to be only that of an 'introducer of information' which I hoped, at the very least, would provide reason for the reader to open their awareness to questionable aspects of so-called 'settled astro-science theory', and at the very best, it might turn out to be an introduction for some to a whole new way of thinking about our universe. As individuals, keen for whatever reason to be involved in the discovery and understanding of accurate knowledge, we have a duty to make ourselves aware of what is going on across the range of scientific research activities that interest us. If we do not do this, then we are in danger of suspending our ability to judge by merely adopting the opinion of those who appear, for some reason, to deserve our respect. Going down this path will not help us develop our own legitimate viewpoints and knowledge; instead, we will end up collecting what is abundant out there, the opinions, prejudices and misconceptions of others.

The simple thing for objectors to do is to dismiss EU supporters as misguided and/or somehow disgruntled folk. However, the smart among them who are already suspicious of what the mainstream has been saying have started on their own to look at alternative theories to refine their core beliefs. Looking at popular astro-science internet forums and previously non-involved science and astronomy websites where discussions on EU-related subjects now take place, it seems that things are changing. Of course there are those who will say the EU is all nonsense, dismiss it and move on. I regret this because I believe one should be open to all types of constructive discussion that lead to progress and I certainly have time for people who make a respectful effort to explain why they think in a particular way. This scenario would also be good for the interested public because it would help maintain a focus, encourage deeper interest and develop a base of understanding on which to build. At the end of the day, however, it will be the real experts in these fields who will present the professional case for change to open-minded scientists with influence to bring that change about. So, to summarise what has been covered ...

Today's Standard Model of the Universe has been inspected and analysed to the point where we now have a good understanding of how and why it has evolved into the confusion of questionable theories we now have. Many of these are shown to either be unworkable in terms of the science on which they are based, or to have arisen with no value solely from wishful, imaginative thinking. The Standard Model is therefore fatally flawed. This state of affairs is generally known but for various reasons it is not being addressed. So for everyone's sake, but especially that of our children, we have a situation in astro-science that cannot be left unchallenged.

It seems clear now that the relevance of fundamental electrical, magnetic and plasma theories in the astro-science research context has been ignored for over a century. Instead, Newtonian mechanics and relativistic gravity theories have been focussed on that have mainly produced uncoordinated and limited useful results.

Modern advancements in technology, general science and in astro-science in particular, have led to more revealing analyses of the better data to which we now have access. Due to the substantial background support for the EU model that has arisen from this, and due to the ever more obvious logic of the situation, we now face the challenge of coming to terms with the realisation that electric currents do indeed flow throughout space.

Electricity requires a conducting path to flow along, and in space those paths are formed from the plasma that we know already exists there. The challenge therefore is to properly understand plasma's ability not to instantly mix to become electrically neutral, as we are told it does, but instead to form in regions that have isolating double-layer barriers between them that allow powerful charge differences to build up. Where these different electrical potentials exist then conditions are right for currents to flow, and when they do flow, magnetic fields are produced to surround and spread out from the paths those currents take. These paths are what we often now visibly see as twisted filaments in space - they are Birkeland currents.

The acknowledgement by astro-science of the reality of Birkeland currents, despite only being a tailored admission to suit its own aspirations, is actually an unintended recognition that electric currents do flow in space, and as we have discussed, this is the case at every conceivable scale. We know that the fundamental electromagnetic force is $10^{39}$ times more powerful than the force of gravity and that its effects on plasma are shown from laboratory experiment to be upwardly scalable by $10^{14}$ times. What results, therefore, is a situation where ionised matter is being collected together in concentrations that span vast distances, the dimensions of which we can only attempt to imagine. These regions of concentrated plasma subsequently interact with each other dynamically through electromagnetic forces, the power of which we can only wonder at and so represent with numbers that have many trailing zeros. The network of Birkeland currents that therefore exists throughout our universe as the cosmic power distribution system, supplies energy to galaxies, stars and all the other energetic phenomena we observe. All of these, fundamentally, are various plasma structures that form and re-form within a cosmic ocean of ionised dust and gas to give us the fabric, objects and events we observe that together present to us our powerfully dynamic but apparently motionless universe. Within all this, and closer to our human scale of appreciation, the relatively minuscule bodies of planets and moons are moved around by gravity, that is yet to be fully understood. It is only when these bodies approach each other that electrical forces dominate and they interact spectacularly by discharge through the plasma medium of space that surrounds them. Happily, it seems that such electrical exchanges serve to change orbits in such a way as to restore peace among the planets.

Keeping the scalability of plasma energy and behaviour in mind, we find evidence of space being electric right in front of our eyes when we look at how comets visually appear and how they behave. These occasional visitors to the inner region of the solar system may open the door for us to everything else electric. The tantalising data now collected around them may turn out to be the seed from which a breakthrough will sprout as evidence to the electrical nature of our universe. Comets are solid bodies isolated electrically from their surroundings by a double layer barrier formed by their own charge level and that of their oppositely charged environment. For most of their long elliptical journey far from the Sun they are negatively charged bodies, so their spectacular visual and physical behaviours as they fly through the rapidly increasing positive charge that surrounds the Sun can be accounted for as resulting from the inevitable electric discharges that take place. It is also during this inner solar system journey that the fracturing, cratering, sculpting and erosion of comets takes place.

Given that these electrical processes hold true at all scales, we then consider them in terms of solid planets and moons. You can imagine similar events happening to these bodies if they experience rapid changes in the charge density that surrounds them, such as where a planet in a short-lived elongated orbit would pass through a gradient of charge density around a star. Remember here that it has been recorded in ancient times that Venus was seen in the sky as a comet with a tail! Thankfully, things in our own solar system now seem peaceful. However, if this electrical balance was to be disturbed, then we could expect spectacular visual displays, and quite possibly, some degree of geological upheaval here on Earth.

Our star, the Sun, does not operate through the action of a gravity-moderated nuclear explosion. It works instead as a concentrated ball of electrically excited plasma, where its visible glowing surface is maintained in arc mode by a drift of electrons flowing inward from the distant cathode region of our solar system, the heliopause, and by positive currents flowing into its poles from the interstellar circuit of which our star is a single component. The Sun is, in fact, an enormous fluorescent light that glows around a smaller, solid and much cooler anode core.

Stars, as concentrations of plasma with their own heliospheres, move around and interact over amounts of time that mean little to us, for the human race has only been around for a hint of a tick of the universal clock. This is why, in order to gain anything close to a relevant perspective on these things, it is important to understand the scale of what is involved; such things as electromagnetic energy, amounts of matter, distance and time. Unless we get our heads around these, we cannot even begin to appreciate the relationship we have with the cosmos.

Stars and regions of ionised dust and gas of all densities and dimensions interact electrically to provide the wondrous phenomena to which individual identities and explanations have traditionally been assigned, none of which are real or relevant. Black Holes, Neutron Stars, Pulsars, Dark Matter, Dark Energy and all other inventions of today's astro-science are, I suggest, explicable by electric current flow through and between regions of ionised matter on scales beyond our ken, and the electromagnetic forces involved in this process.

Electrical stresses grow and diminish continually. Where this occurs in the extreme with stars, they can split apart or eject giant gas and solid planets, these then being manoeuvred to locations where an electrical balance is achieved within their charged environment. Gas giant planets under electrical stress go through this same process to produce the smaller solid bodies of moons that we now find so numerous in orbit around their parents. Every electromagnetic system, such as exists in our solar system, seeks to establish and maintain balance.

The paths along which electric currents flow through ionised matter are surrounded by their self-produced magnetic fields. These fields have direction and keep together the plasma filaments that are subsequently formed in entwined bifilar pairs that have an overall rotational force. These Birkeland currents exist at every scale. They are responsible for the formation and operation of all major bodies in the plasma environment of space through their powerfully attractive, rotating electromagnetic forces and they provide the explanation for why galaxies, stars, planets and moons all display their own natural rotation.

EU theories, though profound in themselves with their implications, do not attempt to address the same array of cosmic subjects and their associated implications that the Standard Model of the universe does. The Electric Universe model is therefore not a replacement for the Standard Model. Its sole purpose is instead to help provide a fresh foundation on which to build a better set of theories about our universe, for this is a subject area that will never disappear as something for humans to wonder about. So what are we to make of the current state of affairs?

The history of science tells us much about how we got here and why things in astro-science still persist to hold back our genuine advancement. Apart from some lessons for the future, there remains little to gain by going over again and again the reasons why this is so, assigning blame and attempting to hold irredeemable people and organisations to account. Rather, I think that we, and here I have in mind a loose cooperation between open-minded scientists, academics (including philosophers) and capable lay researchers, should be attempting to go around the monolith that has evolved by making independent efforts to understand what we already have that is of use, broadening our scope of research and cooperation, forming interdisciplinary plans to forge ahead, and getting on with doing just that.

While appreciating the organisational challenges in this and that access to resources, facilities and data would be a problem, I see the way ahead involving two general areas; one of well supported, organised research by genuine practitioners of interdisciplinary science and one that involves the evermore enlightened and capable public who want to support this inclusive approach. Personally, I am not able to take part in the science but I can certainly help with getting good science information out to the public arena. If looking forward with energy and enthusiasm is therefore the thing to do, what is it that the interested lay public can do to help?

To start, I believe we need to get a little bit emotional about prevailing bad science and understand that any motivation for bringing about change we come to feel, once possessed, cannot be denied. This motivation would be fuel for us as our awareness opens up, our studies intensify and the buds of new knowledge begin to break soil. At the same time we need to understand the absolute need to make up our own minds as individuals about these things and not just thoughtlessly accept that change is needed because others say so. Therefore, for people's efforts to be of most use, their self-education should first involve getting to know 'what' is currently wrong and 'why' certain things need to change.

One bottom line is that the interested public needs to talk more about these things and not just leave opinion forming to such as those who prefer to operate as informed loners, these often being the academically capable in science and math who believe that the person on the street has no useful part to play anyway. Accountable scientists and their work should not be allowed to function as islands of activity. Intellectuals are often not as 'worldly clever' as they would like to think, and in many cases they are prone to having blind spots to things outside their own fields, which by definition they are unaware of. There are, of course, many notable exceptions to this, and these folks are obvious when one looks their work. (I like this article on the subject [7-1].)

We need to find the more worthy scientists and their work because these people are not promoted through everyday channels, and we should also find ways to show our support for whatever aspects of their work we come to understand and agree with. These folk are humans just like the rest of us, and where appropriate, it would be helpful for them to know that their work is being followed with interest. Just think what an injection of enthusiastic interest from the public would do for non-mainstream, interdisciplinary scientists and researchers who through their independent efforts might just be on track to achieve good things for the good of all.

It would also be a good thing for the interested public to develop a long-term interest in science and where it is headed in general. The public has been somewhat lazy and accepting in the past, and within that laziness and acceptance can be found a contribution to why things are as they are today. What has been allowed to go on is not right, it needs to be fixed and those members of the public who are motivated, and who can, have a part to play in doing something about it. So this is what I am asking, when we can generate opportunities or when they come along for us to add constructively to what is going on, and here we have been discussing astro-science, then we need to follow up on these opportunities with a level of cautious and respectful confidence.

My big view of the universe is that, however old it may be, the fact that it functions electrically serves as no clue whatsoever to many of the major things that have been claimed about it, things such as its age, size and shape, whether it is getting bigger or contracting, whether multiple universes exist, or indeed that what we think we know about it that provides a basis to assess the possibility of life existing elsewhere. We are currently in no position to say anything for sure about any of these things, except perhaps the existence of extraterrestrial intelligent life, but the evidence for that being true comes from other areas.

I also believe that the four fundamental forces; the weak and strong nuclear forces, electromagnetism and gravity, will all eventually be seen as having their common root in the electromagnetic force. The range of effect that force has, extends from the tiniest atomic to the largest cosmic limits, where electrical and magnetic polarity alignments and associations between particles and forces with their powerful interactions are the factors that will eventually, I venture, be found as the basis on which to explain all forms of energy and structure. The continuity of explanations that arise from the electromagnetic force's scalability is supported by observation and certified data, where the clarity this brings to the puzzle of how our universe works, is profound.

As has been said, the Electric Universe model is not suggested as a replacement for all that is there today. The contribution it offers would however complement existing good science so that a better model could be produced. The apparent ‘us versus them’ situation is therefore a red herring; any adversarial attitudes perceived to exist are only real within the minds of those who intransigently believe they are right and *whose ears cannot hear because their mouths are so full of words* (old American Indian saying). It is instead cooperation, not confrontation, that is sought as the proper way forward. The task therefore is one of education for open minds in the hope that changes to individual and collective points of view can serve as the doors to better science.

And what about the role of the media? Where adequate money and resources are available, highly polished and very convincing visual presentations of how our universe is said to work have been produced. This is actually a form of information poison! These beguiling visual wonders are most often presented as educational television documentaries, narrated by big names in science who temporarily find themselves in the public limelight, or by famous people from the film industry who do not need to seek the public's attention, they just get it. Impressively effective productions they may be, but we need to see them as just that and dealing with fiction, and that some of them are riddled with words and phrases that turn out to be blindly adopted metaphors that we accept as part of normal speech, such as with ... "the black hole in our finances" or "the whole thing has gone supernova"!

Conveniently in addition, and ironically, we support the super-duper ideas about gravity because of our everyday experience of it. To our minds it is the magical, invisible force that we just accept is there as the dominant one because it lets us have a normal day by keeping our tea in our cups, our cars on the road and our feet on the ground. We are so accepting of the effects of good old gravity in our lives that when a well known persona promotes the gravity based ideas of the Standard Model, we just accept that what we are being told is correct, after all, is it not more convenient to assign an explanation for difficult stuff to something we cannot see? And it is definitely ironic that these amazing documentaries would be so much more wonderful, educational and relevant if the public were to experience them based around the electrical and plasma science theories that I and many others believe are really at the heart of the matter.

Electric Universe supporters and their efforts to bring good information to the attention of open-minded science and the broader public are at a distinct disadvantage here, for funds and resources to promote the EU model currently only come from the contributions of private individuals, rather than from any guaranteed source that is supported by public taxation, as is typically the case with mainstream astro-science. Nothing major in this world seems to change without the prospect of making money off it, and the commercial interests that have grown up around servicing Standard Model theory research are unfortunately substantial in the extreme. Cambridge University's biologist and author Dr Rupert Sheldrake suggested that 1% of today's scientific research funding should be diverted to those who have convincing theories developed outside mainstream science. This funding would be allocated by a board of decision makers consisting of a mix of broad experience rather than all being from science. 1% does not sound like much but in real terms it is an awful lot of money that could make a very big impact. Even though this proposal seemed fair in the eyes of many, it has been pretty much ignored, and I would suggest this has been because it is seen a threat to the status quo.

Considering this book in broad terms, I am prepared for my words to be viewed by some within the 'gravity community' to be non-science and possibly non-sense. Well, some of them would say that, wouldn't they?! To my mind the onus would be on those people to do the work to write the book that describes and justifies their own view in opposition and not just a narrow one handed down to them that they have spent little time thinking about. If anyone did embark on that task then it would be useful for them to keep in mind that the single most important reason behind me writing this book is that, instead of being based solely on ideas and gymnastic conjecture, it comes from a base of common sense science that has its foundations in proven working theory. It is abundantly clear that Electric Universe theories have a far better chance of making sense to fair and open-minded members of the public. But again and of course, there will always be those who believe they already know the truth and that discussion around other ideas is pointless. I have said it before but I'll say it yet again; I just want people to be interested enough to look deeper and to make up their own minds on these things. After all, should we think for ourselves or just wait to be given our opinions?

Perhaps supporters of the Standard Model who have the capacity to accept that there may be something to pay attention to in the Electric Universe model will see this as an intellectual challenge for them to take some time over. After all, looking a little further should hold no fear for these people if they are confident of their knowledge and open-minded. My appeal to them would be to start off by putting aside their normal beliefs before they investigate, consider and discuss Electric Universe theories. Laying things out on the table is always good to do. Nothing can stop public debate if we motivate ourselves to have it and those of us who can should talk openly about areas of disagreement. I cannot be clearer when I say that it is wrong to wait for change to come down from the 'gods' of astro-science's Mount Olympus; we should instead go straight to the interested public who may well end up putting a fire to the toes of the antiques who populate that place by making it obvious that the power of informed public opinion will likely help make them appear as dots in the rear-view mirror of the bus that is headed towards reality.

Thinking about constructive dialogue in science reminds me of something I read in Stephen Hawking's "The Illustrated Brief History of Time" (p94). For me, one thing his words expose is the attitude to change that exists in the upper echelons of astro-science … "*It is very difficult to make a mark in experimental physics unless you are already at the top!*"

I have had the pleasure of being in the company of Wal Thornhill at two public events but I cannot remember whether it was at one of these or somewhere else where he said the following … "*The information I could give to the world's young folks would be just the thing to inspire them in science in ways that have disappeared from current science.*" This, for me, says a lot about the type of people who are involved with the Electric Universe project and the motivations they have for doing what they do. It also highlights the unavoidable responsibility that we as supposed grown-ups have for attending to the wider learning environment so that the minds of our children can expand and develop appropriately into the future. I also liked Wal's summary description of the time when Electric Universe ideas will be seen as the truth of the matter; he referred to this as … "*childhood's end*" … the time when we surrender our arrogance and humbly start understanding the reality of our place in the universe, how it operates, and how the human race should develop into the future. I do have faith that time itself can change all these things. The antiquated old codgers will die off, worthy young brains will take over and the public will feel more engaged with natural science.

One other thing I must say about the attitude and commitment of the people I have had the pleasure of communicating with while journeying into the basics of the Electric Universe theories. What I have found seems to exemplify what I would expect from intelligent broad-minded researchers who have a heart-felt understanding of what making a positive contribution is really about. The difference between what we currently experience from the mainstream and the intellectual glow that these fine folk exude is stark; they have time for every reasonable idea. I only need recall communications with EU supporters to remember seeing words frequently popping up, such as: possible, achievable, potential, can, doable, might, may, feasible, perhaps, likely and workable. These people have open minds and a can-do approach that I have faith will win every time.

There is a very important thing that I would underline as you consider what I have written ... Please do try to rely on your own judgement in these matters by not letting yourself be swayed by self-imposed confidence issues that may quickly send you off seeking the opinion of people whom you think already have a good grasp of these matters for they are precisely the people not to go to first. What I have invariably found is that certain people, those who already have strong opinions on these matters, actually have a narrow viewpoint that seriously blinds them from other possibilities. If you rush off to talk to your friend who you believe already knows a bit about space, it is sadly almost predictable that they will say EU theories are a load of tosh, even before listening to any amount of detail you might want to present. This is the closed-minded thinking I have been banging on about, and I would earnestly encourage you to think this stuff out for yourself before discussing it too openly.

Claiming that something just feels right is simply not acceptable to dyed-in-the-wool scientists and mathematicians because for most of them, a feeling cannot be measured. Here, I know they miss the very important and relevant point that 'intuition' also has a part to play. To judge and maintain good direction, in addition to the rigorous proof that can come from following the scientific method, there is also the often hard to describe element of 'informed intuition' to consider. Intuition once played a major role when science was known and practiced as natural philosophy. Since that time the mind of the mathematician has been sadly promoted as omnipotent and we seem to have forgotten how important informed intuition can be to the healthy development of ideas.

I take full responsibility for any mistakes you may have encountered in this book and I stand in full support of the imperfect but nonetheless correctly aligned Electric Universe model as one of its most dedicated supporters. I also hold my hands up to probably being short of the 100% mark with some the technical descriptions I have documented. I would therefore appeal to those good people whose first reaction is to criticise for the sake of their view of 'good science', to remember that this book is meant for the interested lay reader, especially those who have never read anything of this nature before; it is intended only to get these folks started on their own journey of discovery. My book is not perfect; what or who is? So, if feedback is deemed necessary, please let it be of a constructive nature and not just an attack on what I have tried to do. I can assure everyone that I am open to the constructive comments of those who definitely know better than me, for I am very aware that I can improve this information that I feel so strongly should be shared.

And finally ... Why should any of this matter at all? Well, I suppose I would need to go back to what I feel myself in terms of taking a wider view of who we humans really are, why we are here, where things might be going and our connection with Nature. I believe we have previously known but forgotten about many of these things, and I feel it is time for those of us who are ready, to expand and again practice greater consideration of them. So, in ending, this has been my own little contribution to help bring this about.

**A Final Indulgence ...**

Here is an extract from the book 'Consciousness Beyond Life' by Pim van Lommel, MD [7-2] ...

"*According to the philosopher of science Ilja Maso, most scientists employ the scientific method based on materialist, mechanistic, and reductionist assumptions. It attracts most of the funding, achieves the most striking results, and is thought to employ the brightest minds. The more a vision deviates from its materialist paradigm, the lower its status and the less money it receives. Indeed, experience shows us that the upper echelons of the research hierarchy receive a disproportionate percentage of funding, whereas the lower echelons actually address the conditions, needs and problems of the people. True science does not restrict itself to material and therefore restrictive hypotheses but is open to new and initially inexplicable findings and welcomes the challenge of finding explanatory theories. Maso speaks of an 'inclusive science', which can accommodate ideas that are more compatible with our attempts to learn about subjective aspects of the world and ourselves than the materialist demarcation currently allows.*" [7-3] [and further on in that book ...]

The psychologist Abraham H Maslow offered a fine definition of what such an inclusive science should entail:

"*The acceptance of the obligation to acknowledge and describe all of reality, all that exists everything that is the case. Before all else science must be comprehensive and all-inclusive. It must accept within its jurisdiction even that which it cannot understand or explain, that for which no theory exists, that which cannot be measured, predicted, controlled or ordered. It must accept even contradictions and illogicalities and mysteries, the vague, the ambiguous, the archaic, the unconscious, and all other aspects of existence that are difficult to communicate. At best it is completely open and excludes nothing. It has no entrance requirements.*" [7-4]

# References in the book

## Introduction References:


[I-1] Wikipedia article - “Standard Model”
http://en.wikipedia.org/wiki/The_Standard_Model

[I-2] YouTube video - “Thunderbolts of The Gods”
http://www.youtube.com/watch?v=P4zixnWeE8A


## Chapter One References:


[1-1] NASA web article - “Big Bang Cosmology”
http://map.gsfc.nasa.gov/universe/bb_theory.html

[1-2A] NASA web article - “What is the Universe Made Of?”
http://map.gsfc.nasa.gov/universe/uni_matter.html

[1-2B] NASA web article - “Dark Energy, Dark Matter”
http://science.nasa.gov/astrophysics/focus-areas/what-is-dark-energy/

[1-3] NASA web article - “What is the Inflation Theory?”
http://map.gsfc.nasa.gov/universe/bb_cosmo_infl.html

[1-4] NASA web article - “When Did the First Cosmic Structures Form?”
http://wmap.gsfc.nasa.gov/universe/rel_firstobjs.html

[1-5] Harvard-Smithsonian web article - “White Dwarfs & Planetary Nebulas”
http://chandra.harvard.edu/xray_sources/white_dwarfs.html

[1-6] Harvard-Smithsonian web article - “Neutron Stars / X-ray Binaries”
http://chandra.harvard.edu/xray_sources/neutron_stars.html

[1-7] Harvard-Smithsonian web article - “Black Holes”
http://chandra.harvard.edu/xray_sources/blackholes.html

[1-8] Harvard-Smithsonian web article - “Supernovas & Supernova Remnants”
http://chandra.harvard.edu/xray_sources/supernovas.html

[1-9] Wikipedia article - “Sun”
http://en.wikipedia.org/wiki/Sun

[1-10] Wikipedia article - “Redshift”
http://en.wikipedia.org/wiki/Redshift

[1-11] NASA web article - “How Fast is the Universe Expanding?”
http://wmap.gsfc.nasa.gov/universe/uni_expansion.html

[1-12] Wikipedia article - “Comet”
http://en.wikipedia.org/wiki/Comets

[1-13] Wikipedia article - “Impact Crater”
http://en.wikipedia.org/wiki/Impact_crater

[1-14] NASA web article - “A Lunner Rille”
http://apod.nasa.gov/apod/ap021029.html

[1-15] NASA web article - “Solar System: Sun, Moon and Earth”
http://www.nasa.gov/mission_pages/GLAST/science/solar_system.html

[1-16] Harvard-Smithsonian web article - “Cosmology / Deep Fields / X-ray Background”
http://chandra.harvard.edu/xray_sources/background.html

[1-17] Harvard-Smithsonian web article - “Gamma Ray Bursts”
http://chandra.harvard.edu/xray_sources/grb.html

[1-18] Wikipedia article - “Pioneer Anomaly”
http://en.wikipedia.org/wiki/Pioneer_anomoly

[1-19A] Wikipedia article - “Enceladus (moon)”
http://en.wikipedia.org/wiki/Enceladus_%28moon%29

[1-19B] NASA web article - “Cassini Finds Warm Cracks on Enceladus”
http://www.jpl.nasa.gov/news/news.cfm?release=2010-402

[1-20] NASA web article - “Cassini Images Bizarre Hexagon on Saturn”
http://www.jpl.nasa.gov/news/news.cfm?release=2007-034

[1-21] CERN web article - “The Large Hadron Collider”
http://public.web.cern.ch/public/en/lhc/lhc-en.html

## Chapter Two References:

[2-1] Dr Gerald Pollack Paper -
“Revitalizing Science in a Risk-Averse Culture: Reflections on the Syndrome and Prescriptions for its Cure”
http://tinyurl.com/847r675

## Chapter Three References:

[3-1] Keck Observatory website article – “NASA Study Distinguishes Most Distant Galaxy Cluster”
http://keckobservatory.org/news/nasa_study_distinguishes_most_distant_galaxy_cluster1/

[3-2] Wallace Thornhill website article – “A Real Theory of Everything”
http://www.holoscience.com/news.php?article=gdaqg8df

[3-3] Wallace Thornhill website article – “Deep Impact 2”
http://www.holoscience.com/news.php?article=nq9zna2m

[3-4] Stanford University website article – “GP-B Status Update - May 4, 2011”
http://einstein.stanford.edu/highlights/status1.html

[3-5] Caltech website article - "LIGO Data Management Plan, January 2012"
https://dcc.ligo.org/public/0009/M1000066/017/LIGO-M1000066-v17.pdf

[3-6] Berkeley website article - "CDMSII Overview"
http://cdms.berkeley.edu/experiment.html

[3-7] NASA website article - "Wilkinson Microwave Anisotropy Probe"
http://map.gsfc.nasa.gov

[3-8] Universe Today website article - "Distant Galaxy is Too Massive For Current Theories"
http://www.universetoday.com/10974/distant-galaxy-is-too-massive-for-current-theories/

[3-9] Space Telescope Science Institute website article – "Hubble Detects Faster Than Light Motion in Galaxy M87"
http://www.stsci.edu/ftp/science/m87/press.txt

[3-10] TPOD - "Another electrical 'shock' for Astronomers" - November 13th 2006
http://www.thunderbolts.info/tpod/2006/arch06/061113shockwave.htm

[3-11] Stephen Crothers website - "The Black Hole, the Big Bang and Modern Physics"
http://www.sjcrothers.plasmaresources.com/

[3-12a] Review by Prof. Myron W. Evans of Jeremy Dunning-Davies' book "Exploding a Myth"
http://aias.us/index.php?goto=showPageByTitle&pageTitle=Book_Review:_Exploding_a_Myth_by_Jeremy_Dunning-Davies

[3-12b] Amazon Bookshop – "Giant Galaxy String Defies Models of How Universe Evolved"
http://www.amazon.co.uk/Exploding-Myth-Conventional-Wisdom-Scientific/dp/1904275303

[3-13] New Scientist website article - "Baby Star Found Near Galaxy's Violent Centre"
http://www.newscientist.com/article/dn9738-baby-star-found-near-galaxys-violent-centre.html

[3-14] Wikipedia article - "Pioneer Anomaly"
http://en.wikipedia.org/wiki/Pioneer_anomoly

[3-15] Universe Today website article – "The Sound of Saturn's Rings"
http://www.universetoday.com/51511/the-sound-of-saturns-rings/

[3-16] Physics website article - "Free Fall of Elementary Particles"
http://www.electrogravityphysics.com/html/contents.html

[3-17] General website discussion article – "The Impossible Dinosaurs - Megafauna and Attenuated Gravity"
http://www.freerepublic.com/focus/f-chat/1989265/posts

## Chapter Four References:

Because this is intended as a general information chapter, I have not supplied any references within it for the reader. I do not want to be accused of being selective with the fundamental information it should deal with by pointing to sources that could be seen as supportive in an undue way for what I go on to say in the rest of the book. Therefore, if questions arise or interest is piqued, the reader should look for themselves for standard explanations that go further. Using the internet is the most convenient way to do this, but I would suggest sticking to the major educational websites.

## Chapter Five References:

[5-1] Kristian Birkeland - http://www.plasma-universe.com/Kristian_Birkeland

[5-2] Irving Langmuir - http://www.plasma-universe.com/Irving_Langmuir

[5-3] Hannes Alfvén - http://www.plasma-universe.com/Hannes_Alfven

[5-4] Immanuel Velikovsky - http://www.velikovsky.info/Immanuel_Velikovsky

[5-5] Charles E R Bruce - http://www.plasma-universe.com/Charles_Bruce

[5-6] Ralph Juergens - http://www.velikovsky.info/Ralph_Juergens

[5-7] Halton Arp - http://www.haltonarp.com/

[5-8] Anthony Peratt - http://www.cambridgewhoswho.com/Images/Site/DocumentManager/Anthony%20Peratt%20Biography.pdf

[5-9a] David Talbott - http://www.velikovsky.info/David_Talbott

[5-9a] Natural Philosophy Alliance - http://www.worldnpa.org/site/

[5-10] Wallace Thornhill - http://www.velikovsky.info/Wallace_Thornhill

[5-11] Donald E Scott - http://www.mikamar.biz/book-info/tes-a.htm

## Chapter Six References:

**NOTE**: TPODs are Thunderbolts Picture of the Day articles on the www.thunderbolts.info website. The chronological TPOD archive from July 2004 to September 2011 can be found here: http://www.thunderbolts.info/tpod/00archive.htm TPOD articles published since that time period are accessible through the website's home page.

[6-1] Web article - "Electric Currents and Transmission Lines in Space"
http://public.lanl.gov/alp/plasma/elec_currents.html

[6-2] Tom Wilson, TPOD - "A New Look at Near Neighbors Part One" - October 21st 2009
http://www.thunderbolts.info/tpod/2009/arch09/091021neighbors.htm

[6-3] Stephen Smith, TPOD - "Presumptive Prepolyds" - June 13th 2010
http://www.thunderbolts.info/tpod/2010/arch10/100713proplyds.htm

[6-4] Stephen Smith, TPOD - "Luminous nebulae confirm Electric Universe theory" - April 18th 2012
http://www.thunderbolts.info/wp/2012/04/17/spider-bites/

[6-5] Stephen Smith, TPOD - "Stars That Will Not Explode" - June 19th 2012
http://www.thunderbolts.info/wp/2012/06/19/stars-that-will-not-explode/

[6-6] Stephen Smith, TPOD - "A Mystifying Menagerie" - February 4th 2011
http://www.thunderbolts.info/tpod/2011/arch11/110204menagerie.htm

[6-7a] TPOD - "The Iron Sun Debate (4)" - January 26th 2006
http://www.thunderbolts.info/tpod/2006/arch06/060126solar4.htm

[6-7b] Wal Thornhill, "Our Misunderstood Sun" (2010)
http://www.holoscience.com/news.php?article=ah63dzac

[6-8] TPOD - "Saturn's Surprises Will Point to Electrical Origins " - December 28th 2004
http://www.thunderbolts.info/tpod/2004/arch/041228prediction-origins.htm

[6-9a] Anthony Watts, - "Deep Purple Haze – the orginal sunscreen"
http://wattsupwiththat.com/2010/06/04/deep-purple-haze-the-orginal-sunscreen/

[6-9b] TPOD - "Predictions Concerning Titan's Methane" - August 4th 2006
http://www.thunderbolts.info/tpod/2006/arch06/060804titansmethane.htm

[6-10] Wal Thornhill, "Planet Birthing" (2003)
http://www.holoscience.com/news.php?article=rbkq9dj2

[6-11] Mel Acheson, TPOD - "Stars in Collision Part 1" - May 20th 2010
http://www.thunderbolts.info/tpod/2010/arch10/100520collision1.htm

[6-12] Mel Acheson, TPOD - "Stars in Collision Part 2" - May 21st 2010
http://www.thunderbolts.info/tpod/2010/arch10/100521collision2.htm

[6-13] TPOD - "Saturn in Ancient Times" - September 23rd 2004
http://www.thunderbolts.info/tpod/2004/arch/040923saturn-ancient.htm

[6-14] Stephen Smith, TPOD - "Solar Plasma Circuits" - March 22nd 2010
http://www.thunderbolts.info/tpod/2010/arch10/100322circuits.htm

[6-15a] Wal Thornhill, "Newton's Electric Clockwork Solar System" (2009)
http://www.holoscience.com/news.php?article=q1q6sz2s

[6-15b] Wal Thornhill, "Electric Gravity in an Electric Universe" (2008)
http://www.holoscience.com/news.php?article=89xdcmfs

[6-16] Michael Armstrong, TPOD - "A, B, C, D, Electric Solar System" - November 23rd 2006
http://www.thunderbolts.info/tpod/2006/arch06/061123abcd.htm

[6-17] TPOD - "Discovering the Magnetosphere" - September 30th 2004
http://www.thunderbolts.info/tpod/2004/arch/040930magneto-sphere.htm

[6-18] Web document - "In Memoriam Grote Reber 1911-2002 Founder of Radio Astronomy"
http://public.lanl.gov/alp/plasma/downloads/GroteReber.pdf

[6-19] Stephen Smith, TPOD - "Windy Galaxies" - November 1st 2011
http://www.thunderbolts.info/wp/2011/10/31/windy-galaxies/

[6-20] Stephen Smith, TPOD - "It's Birkeland's Birthday" - December 13th 2010
http://www.thunderbolts.info/tpod/2010/arch10/101213birkeland.htm

[6-21] Wal Thornhill, "Twinkle, twinkle electric star" (2008)
http://www.holoscience.com/news.php?article=x49g6gsf

[6-22a] Wal Thornhill, "Failed Star or Failed Science" (2000)
http://www.holoscience.com/news/failed_star.html

[6-22b] Wal Thornhill, "Electric Sun Verified" (2009)
http://www.holoscience.com/news.php?article=74fgmwne

[6-23] Wal Thornhill, “Mystery of the Shrinking Red Star” (2009)
http://www.holoscience.com/news.php?article=jdjcab6s

[6-24] Stephen Smith, TPOD - “Novus Ratio” - June 2nd 2011
http://www.thunderbolts.info/tpod/2011/arch11/110602novus.htm

[6-25] TPOD - “Electric Supernovae” - October 11th 2005
http://www.thunderbolts.info/tpod/2005/arch05/051011elec-nova.htm

[6-26] Stephen Smith, TPOD - “Down the Barrel” - April 11th 2011
http://www.thunderbolts.info/tpod/2011/arch11/110411barrel.htm

[6-27] NASA Web article - “What is meant by false colour?”
http://starchild.gsfc.nasa.gov/docs/StarChild/questions/question20.html

[6-28] Stephen Smith and Jason Brown, TPOD - “Pulsed Power” - May 6th 2010
http://www.thunderbolts.info/tpod/2010/arch10/100506power.htm

[6-29] TPOD - “Neutron Star Refutes It's Own Existence” - July 21st 2006
http://www.thunderbolts.info/tpod/2006/arch06/060721neutronstar.htm

[6-30] Stephen Smith, TPOD - “Magnetic Monsters” - November 26th 2008
http://www.thunderbolts.info/tpod/2008/arch08/081126magnetic.htm

[6-31] TPOD - “Magnetars - A Computer's Dream World” - March 1st 2006
http://www.thunderbolts.info/tpod/2006/arch06/060301magnetar.htm

[6-32a] TPOD - “Prediction #1: Big Bang a Big Loser in 2005” - December 27th 2004
http://thunderbolts.info/tpod/2004/arch/041227prediction-bigbang.htm

[6-32b] TPOD - “The Search for Two Numbers” - September 3rd 2004
http://www.thunderbolts.info/tpod/2004/arch/040903redshift.htm

[6-33] Wal Thornhill - “The Remarkable Slowness of Light” (2002)
http://www.holoscience.com/news/slow_light.html

[6-34] Stephen Smith, TPOD - “Faster Than Light: Part Two” - December 17th 2009
http://www.thunderbolts.info/tpod/2009/arch09/091217light2.htm

[6-35] TPOD - “The Picture that Won't Go Away” - August 31st 2006
http://www.thunderbolts.info/tpod/2006/arch06/060831picture.htm

[6-36] TPOD - “Quasars: Massive or Charged?” - February 12th 2010
http://www.thunderbolts.info/tpod/2010/arch10/100212massive.htm

[6-37] Mel Acheson, TPOD - “Quasar Clusters” - August 17th 2010
http://www.thunderbolts.info/tpod/2010/arch10/100817clusters.htm

[6-38] Web article - “Fundamentals of electricity - Capacitors”
http://www.electronicstheory.com/html/e101-30.htm

[6-39] David Talbott web article - “Comet Elenin—the Debate that Never Happened”
http://www.thunderbolts.info/wp/2011/10/06/comet-elenin%E2%80%94the-debate-that-never-happened/

[6-40] Wal Thornhill, “Deep Impact 2” (2010)
http://www.holoscience.com/news.php?article=nq9zna2m

[6-41] Stephen Smith, TPOD - "Kuiper Belt Objects" - February 17th 2009
http://www.thunderbolts.info/tpod/2009/arch09/090217objects.htm

[6-42] Stephen Smith, TPOD - "A Spectre Haunts the Darkness" - June 1st 2011
http://www.thunderbolts.info/tpod/2011/arch11/110601darkness.htm

[6-43] TPOD - "Deep Impact and Shoemaker-Levy 9" - July 18th 2005
http://www.thunderbolts.info/tpod/2005/arch05/050718deepimpact.htm

[6-44] Stephen Smith, TPOD - "Comet Holmes 17P Startles Astronomers" - October 31st 2007
http://www.thunderbolts.info/tpod/2007/arch07/071031cometholmes.htm

[6-45] Wikipedia article - "Hydroxyl Radical OH"
http://en.wikipedia.org/wiki/Hydroxyl_radical

[6-46] TPOD - "The Comet and the Future of Science" - April 2nd 2007
http://www.thunderbolts.info/tpod/2007/arch07/070402cometfuture.htm

[6-47a] Scott Wall web article - "Comet Holmes - a Media Non-event" - March 29th 2008
http://www.thunderbolts.info/thunderblogs/archives/guests08/032908_guest_swall.htm

[6-47b] TPOD - "When Comets Break Apart" - january 19th 2006
http://www.thunderbolts.info/tpod/2006/arch06/060119comets.htm

[6-48a] Wal Thornhill, "Comet Wild 2" (2004)
http://www.holoscience.com/news.php?article=ayxpdjcb

[6-48b] YouTube video - "Electric Crater Central Buildup"
http://www.youtube.com/watch?v=NaJ3OanxiW8

[6-49] NASA web article - "NASA Spacecraft Finds Comet Has Hot, Dry Surface "
http://www.jpl.nasa.gov/releases/2002/release_2002_80.html

[6-50] Wal Thornhill, "Comet Borrelly rocks core scientific beliefs" (2001)
http://www.holoscience.com/news/comet_borrelly.html

[6-51] Stephen Smith, TPOD - "The Dust of Creeds Outworn" - March 15th 2011
http://www.thunderbolts.info/tpod/2011/arch11/110315dust.htm

[6-52] Mel Acheson, TPOD - "Where Do Asteroids Come From" - May 30th 2011
http://www.thunderbolts.info/tpod/2011/arch11/110530asteroids.htm

[6-53] Stephen Smith, TPOD - "Cometary Asteroids" - May 18th 2011
http://www.thunderbolts.info/tpod/2011/arch11/110518asteroids.htm

[6-54] Wikipedia article - "Allan Hills 84001"
http://en.wikipedia.org/wiki/Allan_Hills_84001

[6-55] TPOD - "Electric Meteorites?" - February 9th 2005
http://www.thunderbolts.info/tpod/2005/arch05/050209meteorite.htm

[6-56] TPOD - "Retrospective on Io" - April 6th 2005
http://www.thunderbolts.info/tpod/2005/arch05/050406retrospective-io.htm

[6-57] NASA web article - "Io's Alien Volcanoes"
http://science.nasa.gov/science-news/science-at-nasa/1999/ast04oct99_1/

[6-58] TPOD - "Meteor Crater in Arizona" - January 31st 2006
http://www.thunderbolts.info/tpod/2006/arch06/060131crater.htm

[6-59] Stephen Smith, TPOD - "Man in the Moon" - March 6th 2006
http://www.thunderbolts.info/tpod/2006/arch06/060305moon.htm

[6-60] TPOD - "Holes in Moons - and in Theories" - June 8th 2007
http://www.thunderbolts.info/tpod/2007/arch07/070608holesinmoons.htm

[6-61] TPOD - "Electric Craters on Planets and Moons" - November 15th 2004
http://www.thunderbolts.info/tpod/2004/arch/041115craters.htm

[6-62] Wal Thornhill, "2008 – Year of the Electric Universe" - News Article Jan.2008
http://www.holoscience.com/wp/2008-year-of-the-electric-universe/

[6-63a] Wal Thornhill, "More on Mercury's Mysteries" (2008)
http://www.holoscience.com/news.php?article=8qysa3zk

[6-63b] Stephen Smith, TPOD - "Son of Zeus" (2011)
http://www.thunderbolts.info/wp/2011/11/15/son-of-zeus/

[6-64] Stephen Smith, TPOD - "Reconnoitring the Moon" - June 19th 2009
http://www.thunderbolts.info/tpod/2009/arch09/090619moon.htm

[6-65] Michael Armstrong, TPOD "Earth's Richart Crater" - January 4th 2006
http://www.thunderbolts.info/tpod/2006/arch06/060104richat.htm

[6-66] TPOD - "Gooches Crater, Australia" - October 8th 2008
http://www.thunderbolts.info/tpod/2008/arch08/081008gooches.htm

[6-67] Stephen Smith and Brad Benson, TPOD - "Lake of the Woods" - December 24th 2008
http://www.thunderbolts.info/tpod/2008/arch08/081224lake.htm

[6-68] Stephen Smith, TPOD - "Kuiper Crater Rays" - April 20th 2009
http://www.thunderbolts.info/tpod/2009/arch09/090420rays.htm

[6-69] TPOD - "The Moon and its Rilles" - March 21st 2006
http://www.thunderbolts.info/tpod/2006/arch06/060321rille.htm

[6-70a] Bert Hickman, web article - "What are Lichtenberg figures, and how do we make them?"
http://www.capturedlightning.com/frames/lichtenbergs.html

[6-70b] TPOD - "The Dendritic Ridges of Olympus Mons" - November 22nd 2006
http://www.thunderbolts.info/tpod/2006/arch06/061122omridges.htm

[6-71] TPOD - "Message of Valles Marineris" - May 16th 2005
http://www.thunderbolts.info/tpod/2005/arch05/050516marineris.htm

[6-72] Wal Thornhill, "Water on Mars?" (2002)
http://www.holoscience.com/news/wateronmars.html

[6-73] Wal Thornhill, "Spiral Galaxies & Grand Canyons" (2003)
http://www.holoscience.com/news.php?article=rnde0zza

[6-74] TPOD - "Etched Mars" - December 27th 2006
http://www.thunderbolts.info/tpod/2006/arch06/061227etchedmars.htm

[6-75] TPOD - "More Strange Lava Tubes of Mars" - November 28th 2005
http://www.thunderbolts.info/tpod/2005/arch05/051128pavonis.htm

[6-76] TPOD - "Mars Bears witness" - September 19th 2005
http://www.thunderbolts.info/tpod/2005/arch05/050919marswitness.htm

[6-77] Stephen Smith, TPOD - "Mars Lights and Lightning" - March 11th 2010
http://www.thunderbolts.info/tpod/2010/arch10/100311lights.htm

[6-78] TPOD - "The Gullies of Russell Crater on Mars" - February 6th 2008
http://www.thunderbolts.info/tpod/2008/arch08/080206gullies.htm

[6-79] Stephen Smith and Thane Hubbell, TPOD - "Nuclear War God" - April 20th 2011
http://www.thunderbolts.info/tpod/2011/arch11/110420wargod.htm

[6-80] TPOD - "Lightning-Scarred Gods and Monsters" - April 20th 2005
http://www.thunderbolts.info/tpod/2005/arch05/050412scarface.htm

[6-81] Michael Steinbacher paper - "A New Approach to Mountain Formation"
http://tinyurl.com/6p5x7pu

[6-82] Andreas Otte's presentation on Michael Steinbacher's work - "A New Approach to Mountain Formation"
http://www.chrono-rekonstruktion.de/mountain-formation/steinbacher.pdf

**Chapter Seven References:**

[7-1] Pravin Singh, Institute of Education, University of the South Pacific, Article in Journal of Educational Studies no.7, 1981
http://directions.usp.ac.fj/collect/direct/index/assoc/D769861.dir/doc.pdf

[7-2] Pim van Lommel MD, website and book "Consciousness Beyond Life"
http://www.pimvanlommel.nl/?home_eng

[7-3] I. Maso, "Arguments in Favour of an Inclusive Science", paper presented at the conference "Science, Worldview and Us", Brussels, Belgium, June 2003.

[7-4] A H Maslow, "The Psychology of Science" (New York: Harper & Row, 1966), chapter 8.

CPSIA information can be obtained at www.ICGtesting.com
Printed in the USA
LVIW01n0011061016
507604LV00007B/19